Reactivity and Structure
Concepts in Organic Chemistry

Volume 17

Editors:
Klaus Hafner   Jean-Marie Lehn
Charles W. Rees   P. von Ragué Schleyer
Barry M. Trost   Rudolf Zahradnik

Robert B. Bates/Craig A. Ogle

Carbanion Chemistry

Springer-Verlag
Berlin Heidelberg New York Tokyo 1983

Robert B. Bates
Craig A. Ogle

Department of Chemistry, University of Arizona
Tucson, AZ 85721/USA

List of Editors

Professor Dr. Klaus Hafner
Institut für Organische Chemie der TH
Petersenstr. 15, D-6100 Darmstadt

Professor Dr. Jean-Marie Lehn
Institut de Chimie, Universite de Strasbourg
1, rue Blaise Pascal, B.P. 296/R8, F-67008 Strasbourg-Cedex

Professor Dr. Charles W. Rees, F. R. S. Hofmann
Professor of Organic Chemistry, Department of Chemistry
Imperial College of Science and Technology
South Kensington, London SW7 2AY, England

Professor Dr. Paul v. R. Schleyer
Lehrstuhl für Organische Chemie der Universität Erlangen-Nürnberg
Henkestr. 42, D-8520 Erlangen

Professor Barry M. Trost
Department of Chemistry, The University of Wisconsin
1101 University Avenue, Madison, Wisconsin 53706, U.S.A.

Prof. Dr. Rudolf Zahradník
Tschechoslowakische Akademie der Wissenschaften
J.-Heyrovský-Institut für Physikal. Chemie und Elektrochemie
Máchova 7, 121 38 Praha 2, C.S.S.R.

ISBN-13:978-3-642-69039-6   e-ISBN-13:978-3-642-69037-2
DOI: 10.1007/978-3-642-69037-2

2152/3020-543210

Preface

This book was prepared with the idea of collecting some of the multitudinous new literature on carbanions and presenting it along with the fundamentals of carbanion chemistry. Some 400 papers from the period 1976 to 1982 have been assimilated into the book along with about an equal number of references (many to reviews) from the earlier literature.

The material is organized under the headings *Structures* (with emphasis on the results of X-ray studies since they unambiguously show the molecular geometry), *Preparations*, and *Reactions*. Reactions, the largest topic, is divided into reactions with electrophiles, eliminations, oxidations, and rearrangements. Under these headings, carbanions without resonance stabilization (σ carbanions) are generally discussed first, and are further subdivided into sp^3 (alkyl) followed by sp^2 (vinyl and aryl) followed by sp (acetylenic); resonance-stabilized (π) hydrocarbon anions such as allyl and benzyl follow; lastly, heteroatom-containing π carbanions such as enolates are considered. Some references to carbanion equivalents are included at the end.

Tucson, Arizona U.S.A.
May 1983

Robert B. Bates
Craig A. Ogle

Table of Contents

I. **Introduction** 1

II. **Structures** 3
   1. Non-delocalized (σ) 3
      a. sp^3 3
      b. sp^2 and sp 7
   2. Delocalized (π) 8
      a. Hydrocarbon Anions 8
      b. Heteroatom-containing Carbanions . . . 10

III. **Preparations** 13
   1. From Alkyl Halides 13
      a. With Metals 13
      b. With Organometallics 15
   2. From Alkanes by Proton Abstraction . . . 17
      a. Acidities of CH Protons 17
      b. Base-solvent Systems 21
   3. From Unsaturated Compounds 25
      a. By Reduction 25
      b. By Addition 26
   4. From Other Organometallics by Changing the
      Metal 27
      a. With Metals 27
      b. With Metal Salts 27

IV. **Reactions of σ Carbanions with Electrophiles** . . . 29
   1. Substitution Reactions of Alkyl. (sp^3) Anions 29
   2. Addition Reactions of Alkyl (sp^3) Anions 34
   3. Vinyl (sp^2), Aryl (sp^2), and Acetylenic (sp)
      Anions 38

V. **Reactions of π Carbanions with Electrophiles** . . . 40
   1. Hydrocarbon π Anions 40
   2. Enolate Anions 43

a. Substitutions 43
b. Additions 45
3. Other O-Stabilized Monoanions 47
4. O-Stabilized Dianions 50
5. N-stabilized Anions 52
6. S-stabilized Anions 56

VI. **Eliminations** 57

1. α-Eliminations 57
2. β-Eliminations 58
3. γ-Eliminations 60
4. δ-Eliminations 61
5. Cycloeliminations 61

VII. **Oxidations** 63

1. Oxidations to Hydroperoxides, Alcohols, and
   Ketones 63
2. Oxidative Couplings 64
3. Dianion Oxidations 65

VIII. **Rearrangements** 67

1. Intermolecular Rearrangements 67
2. Intramolecular Additions 68
3. Intramolecular Eliminations 69
4. Sigmatropic Carbanion Rearrangements . . 70
5. Electrocyclic Carbanion Rearrangements . . 72
6. Complex Intramolecular Rearrangements . . 73

IX. **Carbanion Equivalents** 75

X. **Summary** 77

XI. **References** 78

Subject Index 97

I. Introduction

Carbanions (*1*) are among the most important synthetic organic intermediates, largely because of their ability to form carbon-carbon bonds in high yield by reaction with a great variety of electrophiles, e.g., alkyl halides and ketones:

$$RR' \xleftarrow{\ R'X\ } R^- \xrightarrow{\ R'\overset{\overset{O}{\|}}{C}R''\ } R\underset{\underset{R''}{|}}{\overset{\overset{R'}{|}}{C}}-O^- \qquad R = alkyl$$
$$X = halogen$$

1

Improved methods permitting the preparation of many new types of carbanions paired with many different metals (and even without counterions!) have been responsible for the spectacular growth of this area since Cram's book appeared in 1965 [1]. The Journal of Organometallic Chemistry *weekly* includes many new organometallics, and annually provides surveys of the literature on the organometallics of selected metals.

This book will be limited to relatively "free" carbanions, i.e., those paired with group IA and IIA metals, aluminium, zinc, cadmium, quaternary ammonium ions, and those studied in the gas phase in the absence of counterions. Earlier books in this series discuss organic synthesis by means of transition metal complexes [2, 3].

1

II. Structures

1. Non-delocalized (σ)

a. sp³

That the charge-bearing carbon atom in simple alkyl anions is sp³ rather than sp² hybridized has been amply demonstrated in the gas phase by photoelectron spectroscopy (methyl carbanion without counterion or solvent [4]), in solution by NMR spectroscopy [5, 6], and in crystals by X-ray studies (monomeric EtMgBr · 2 Et₂O [7], dimeric bicyclo[1.1.0]butan-1-yllithium TMEDA [8], tetrameric MeLi [9] and EtLi [10], and hexameric cyclohexyl-lithium · 1/3 benzene [11]).

The state of aggregation (monomeric, dimeric, tetrameric, hexameric, or polymeric) depends on the carbanion structure, cation, temperature, and solvent; coordinating solvents such as ethers and tertiary amines tend to break down the larger aggregates into more reactive [12] smaller ones [13, 14]. That these particular aggregates (with structures given below) should represent energy minima is supported by MO calculations [15].

The organometallic bond may be from a carbon to one metal atom (2, simple σ bond), to two metal atoms (3, "3-center 2-electron" bond), or to three metal atoms (4, "4-center 2-electron" bond); 3 and 4 are considered to contain alkyl groups "bridging" between metal atoms. Electron density at the expected position amidst the four atoms has been observed in a high-quality X-ray study [10]. Carbon-metal bond lengths involving bridging carbons are longer than those for 2-center bonds, as can be seen from the entries for K, Rb, Mg, and Al in Table 1. 4-Center bonds are not noticeably different in length than 3-center bonds (see Li entries). Since C—M bonds involving sp³, sp², and sp hybridized carbon atoms differ very little in length, they have been lumped together in Table 1.

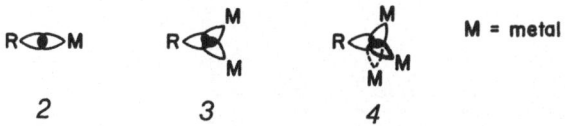

$$
\begin{array}{cccc}
\text{R}\!-\!\text{M} & \text{R}\!-\!\overset{\text{M}}{\underset{\text{M}}{<}} & \text{R}\!-\!\overset{\text{M}}{\underset{\text{M}}{\overset{\text{M}}{<}}} & \text{M = metal} \\
2 & 3 & 4 &
\end{array}
$$

As can be seen in Table 1, the smaller elements (especially Li) are particularly prone to form stable compounds containing bridging alkyl groups.
Monomers — Crystalline EtMgBr · 2 Et₂O [7] and Me₂Mg · TMEDA [36]

3

II. Structures

Table 1. Carbon-Metal Bond Lengths (Å) in Organometallics from X-ray Studies

Metal	Calculated[a]	Observed			
		σ (2-center)	σ (3-center)	σ (4-center)	π (shortest only)[b]
Li	2.00	2.19[14a]	2.21–2.28[8,16]	2.18–2.41[9,10,11,17,18]	2.13–2.34[19–27,27a]
Na	2.34	2.49[28]	—	—	2.64–2.83[29,30]
K	2.80	2.55[28]	—	3.22[31]	3.04–3.16[32,33]
Rb	2.93	2.98[28]	—	3.36[34]	—
Cs	3.12	—	—	3.53[34]	—
Be	1.66	—	1.93[35]	—	—
Mg	2.13	2.09–2.20[7,36–40]	2.22–2.26[40–42]	—	2.26–2.50[40,43]
Ca	2.51				
Sr	2.68				
Ba	2.75				
Al	2.02	1.95–1.99[44,45]	2.24[44]		2.10[45]

[a] Values calculated by adding the covalent radius of carbon (0.77 Å) to the covalent radius of the metal[46]

[b] Only the shortest metal-carbon distance in each π structure is listed

illustrate this group. They both have tetrahedral arrangements of four groups around their magnesium atoms as depicted for the former (*5*) and the latter (*6*), with all bond distances around magnesium being close to calculations from covalent radii.

5   **6**

Dimers — Crystalline bicyclo[1.1.0]butan-1-yllithium · TMEDA [8] and trimethylaluminium [44]; menthyllithium [12] and trimethylaluminium [13] in hydrocarbon solvents illustrate the simplest type. These dimers are held together by two 3-center bonds (*3*) arranged as in *7*; a metal-metal bond is not usually drawn because most of the bond strength is calculated to come from carbon-metal bonding [15]. In the trimethylaluminium dimer, there are two non-bridging methyl groups on each aluminium atom.

8a, a different type of dimer with bridging *halogens*, is observed for EtMgBr · Et$_3$N [37] and EtMgBr · *i*-Pr$_2$O [38]. *8b*, with a chair-shaped six-membered ring, represents a further type of dimer; it is exemplified by 2-lithio-2-methyl-1,3-dithiane · TMEDA (X = sulfur covalently bond to the negatively charged carbon) [27a].

7   **8a**   **8b**

Tetramers — Crystalline MeLi [9], EtLi [10], and MeLi · 1/2 TMEDA [17]; *n*- and *t*-BuLi [13] and Li$_3$MgMe$_5$ [13] in hydrocarbon solvents, and MeLi, *n*-BuLi [47], and Li$_4$Et$_3$OEt [48] in Et$_2$O; EtLi in the gas phase [49]. These are held together by four 4-center bonds (*4*), giving structures *9a* which consist of a tetrahedron of metal atoms with a bridging alkyl group at the center of each face. Direct metal-metal bonds are drawn in *9a* only to aid in visualizing the structure; representation *9b*, in which the atoms strongly bonded to one another are connected, shows that this structure can be thought of as a cube (*9c*) in which 4 non-adjacent corners have been pulled together. Donor atoms such as O, N, S, and P can coordinate with the metal atoms at the tetrahedron corners without disrupting the tetrameric structure [14, 50]; not surprisingly, for steric reasons *t*-butyllithium$_4$ is not solvated appreciably [14]. In the Li$_3$MgMe$_5$ structure, there is a methyl group attached at the magnesium corner of the tetrahedron [13]. Li$_4$Et$_3$OEt presumably has OEt

5

replacing an Et on a face [48]. The stability of these tetramers is illustrated by their survival in the presence of the strongly chelating ligand TMEDA which in $(MeLi)_4(TMEDA)_2$ crystals forms not rings but linear bridge between tetramers via N—Li bonds [17].

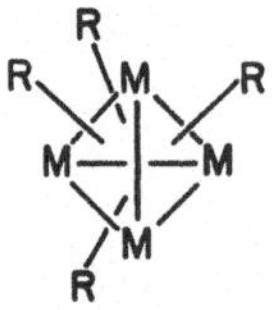

9a 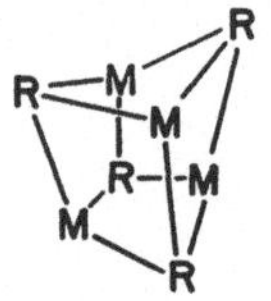 9b 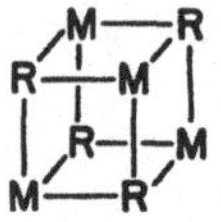 9c

Hexamers — Crystalline cyclohexyllithium$_6$ · benzene$_2$ [11]; EtLi and n-BuL in hydrocarbon solvents [13]; EtLi in the gas phase [49]. The X-ray stud of the former shows these structures to be of type *10*, with a core consistin of a distorted octahedron of metal atoms. Note that two opposite triangula M$_3$ faces (near and far in *10*) bear no alkyl groups; in the X-ray study th Li—Li distances in these triangles were 3.0 Å, whereas the other Li—L distances were only 2.4 Å (compared to 2.4—2.6 Å in tetramers [9, 10, 17] 2.7 Å in a dimer [8], and 2.5 Å in Li metal [46]). These bond lengths help t show that the hexamers are held together by six 4-center bonds (*4*); again the direct metal-metal bonds are drawn for visualization only, and solvatio of the metal atoms at the corners can occur [14].

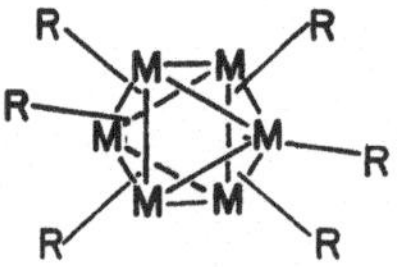

10

Polymers — Crystalline MeK [31], MeRb and MeCs [34], Me$_2$Be [35], Me$_2$M [41], and Et$_2$Mg [42]. The first three are insoluble substances consisting o alternating sheets of alkali metal cations and methyl anions, arranged a in *11* (viewed ⊥ to sheets). Metal atoms in all sheets are directly above on another, but carbanions in every other carbanion layer are displaced s that carbons are very close to the hydrogens of the layer above and below

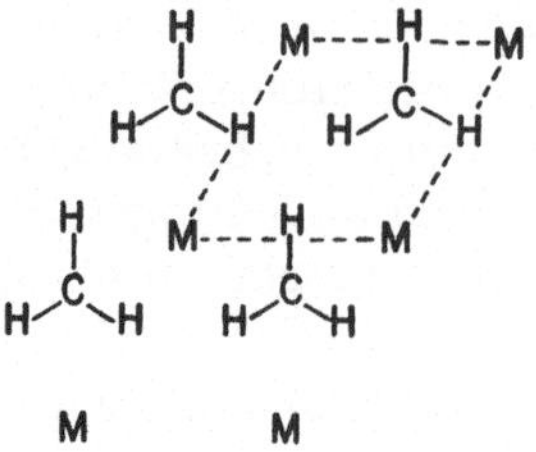

11

The carbons are believed to be sp³ hybridized from IR and weak X-ray evidence; a neutron diffraction study would be helpful. 4-Center bonds (4) are thus apparently used.

The alkaline earth alkyls in this class [35, 41, 42] are linear polymers of the type 12, with a line of metal atoms held together by alkyl groups bridging via 3-center bonds as in certain dimers (7). The alkyl groups are arranged approximately tetrahedrally about the metal atoms.

12

b. sp² and sp

Other carbanions in which the negative charge is not delocalized by resonance over several atoms are the vinyl and aryl anions, with an electron pair localized in an sp² orbital, and cyanide ion and the acetylenic anions, with an electron pair localized in an sp orbital. ΦMgBr · 2 Et$_2$O [39] and · 2 THF [51] are monomeric in the solid state, with structures like EtMgBr · 2 Et$_2$O (5) [39]. Crystalline phenyllithium · TMEDA has a dimeric structure similar to that of the sp³ dimers 4 [16]; presumably the same type of dimer is present in phenyllithium in ether solution [47]. An X-ray study shows (2,6-dimethoxyphenyllithium)$_6$ · Li$_2$O to consist of two approximately tetrahedral Li$_4$ clusters with aryl groups at three faces connected via an oxide ion at the fourth; one such cluster is depicted in 13 [18]. The C—Li bond lengths in this aryllithium dimer and tetramer are similar to those in the alkyllithiums. X-ray studies on ethynylsodium, -potassium, and -rubidium, and 1-propynylsodium and -potassium show metal-carbon distances involving σ-bonding to an sp orbital; there is also some interaction (at a greater distance) between the metal atom and an acetylenic π bond as shown in 14 [28].

13   14

The configurational stabilities of lithium and magnesium salts of alkyl anions (sp³, *optical* isomerization at the charge-bearing carbon) and vinyl anions (sp², *geometric* isomerization about the double bond) have been examined in a few cases, with the general finding of low barriers in the former [5] and high [52, 53] but not insurmountable [54] barriers in the latter. Cyclopropyl anions, which are in between sp² and sp³ in hybridization, have high barriers to inversion [55, 56]. The barriers are generally much higher in non-polar solvents [53]. Some studies of the exchange rates between the various aggregated forms have been made [13].

2. Delocalized (π)

a. Hydrocarbon Anions

Delocalization of negative charge over two or more atoms through resonance as in allyl anion *15* has an enormous stabilizing effect, especially in the gas phase without a counterion [57]. Many X-ray studies on metal salts of such carbanions show them to consist of contact ion pairs, with the carbons in the π systems sp^2 hybridized and coplanar or roughly coplanar, and the metal atoms usually positioned above one or several of the most negatively charged carbons; examples are cyclopentadienylsodium · TMEDA (*16*) [29], cyclopentadienylmagnesium bromide · TMEDA [43], [Me$_5$CpAlMeCl]$_2$ [45], hexatriene dianion dilithium · 2 TMEDA (*17*) [19], benzyllithium · triethylene-diamine *18* [20], 1,3,5,7-tetramethylcyclooctatetraene dianion dipotassium · 2 diglyme *19* [32], indenyllithium · TMEDA [21], bis(indenyl)magnesium [40]. naphthalene dianion dilithium · 2 TMEDA [22], acenaphthene dianion di-lithium · 2 TMEDA [23], fluorenyllithium · 2 quinuclidine [24], fluorenyl-potassium · TMEDA [33], anthracene dianion dilithium · 2 TMEDA [25], triphenylmethyllithium · TMEDA [26], triphenylmethylsodium · TMEDA [30], *20* [27], and phthalocyanine dianion dipotassium · 2 diglyme and · 2(18-crown-6) [58]. The shortest carbon-metal distances (Table 1) in these π-carbanions are greater than for 2-center bonds, and about the same as for 3- and 4-center bonds. The π-carbanions have longer distances than the 3-center bond examples with Mg [40], but the shortest C—Li distance observed so far is in a π case: 2.13 Å in acenaphthene dianion dilithium · 2 TMEDA [23].

UV-visible [59, 60], infrared-Raman [61], and NMR [62–64] studies indicate that in solution contact ion pairs again predominate for metal salts of π-carbanions, except for highly delocalized carbanions (e.g., fluorenyl) which exist as solvent separated ion pairs in strongly cation-solvating solvents like DMSO. The behavior of the carbanion in solvent separated ion pairs is (as expected) cation independent, but in contact ion pairs the geometry, solubility, and reactivity of the carbanion are often very cation dependent. For example, whereas pentadienyllithium (*21*) is mostly W-shaped in THF solution and gives largely reaction products with *trans* double bonds, pentadienylpotassium (*22*) is U-shaped (giving better coordination to the large cation) and gives products with *cis* double bonds [65]. A dimeric cyclohexadienyllithium derivative has been shown by ^{7}Li and ^{13}C NMR to have the "triple ion"-containing sandwich structure *23* [64].

21   *22*   *23*

Certain allylmagnesium compounds apparently prefer covalent σ structures to ion-paired π structures, at least under some conditions [67].

Rotation barriers in allyl and pentadienyl anions in solution are 10–25 Kcal/mole, depending on substituents and conditions [67–69]; rotations about the *inner* C—C bonds in pentadienyl anions equilibrate the U, sickle, and W shapes and are generally faster than rotations about the outer C—C bonds. Alkyl substituents on C-1 of an allyl or pentadienyl group in solution prefer the *cis* position (*24*) [70, 71], presumably because of their hyperconjugative resemblence to butadiene dianion (*25*) [72–74]. The latter dianion has not conclusively been demonstrated to prefer the *cis* shape shown, but many substances with isoelectronic π systems (e.g., *26*, X = halogen) *have*, and dianion *15* has been shown to have this sort of shape in solution [72] as well as in the solid state [19]. The greater stability of the *cis* isomer in all of these cases is presumably due to favorable 1,4 overlap in the HOMO [75, 76]. Surprisingly, in the gas phase without counterions, the *trans* forms of 1-alkylallyl anions appear to be more stable [77].

24   *25*   *26*

Rotation barriers in these π anions (and in their heteroatom-containing analogs discussed in the next section [78]) are higher as the alkali metal size increases [68, 78, 79]; this may be due to the greater ability of the smaller cations to form a σ bond at one carbon long enough to permit rotation about

9

resulting C—C single bonds [79, 80]. Traces of oxygen efficiently catalyze bond rotation in allyl anions, probably by electron transfer to allyl radicals, in which the barriers are much reduced [71].

Monoanions of type (27) with two sp carbons in line with an sp^2 carbon form very readily from allenes and non-terminal acetylenes [81].

Some highly lithiated polymeric species of unknown structure have been prepared which presumably contain some combination of σ and π carbon-lithium bonds [82–84]. An example is C_3Li_4, prepared by perlithiating methylacetylene. It is possible that it contains structure 28, with two σ and two π C—Li bonds and an arrangement of three sp hybridized carbons in a line which is found in "sesquiacetylenic" dianions [85]. By observing their thermal interconversions at 220 °C, the stability order $(C_2Li_2)_n > (C_3Li_4)_n > (C_2Li_4)_n > (CLi_4)_n$ has been established [84]. It would be helpful to have X-ray results on some of these substances to clarify their structures.

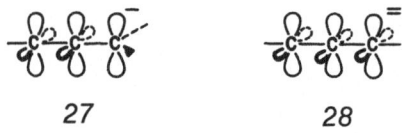

27   28

b. Heteroatom-containing Carbanions

Enolate ions (29) are the most important members of this large class. Although the negative charge is largely on oxygen with resulting high rotation barriers about the adjacent carbon-carbon ~double bond, NMR studies show that *some* negative charge resides on the α-carbon [86–88]. It has been calculated that while monomeric π-bonded allyllithium is more stable than σ-bonded by 16 Kcal/mole, lithium enaminate 30 is only 4 Kcal/mole more stable π-bonded than σ-bonded (Li to N) and lithium enolate (29) is 7 Kcal/mole more stable σ-bonded with the lithium coordinated to oxygen and 176° away from the nearest carbon [89]. It should be noted, however, that enolates are often aggregated, especially in non-polar solvents (see below), and these calculations for monomeric species thus may not be completely relevant. The higher carbon-carbon bond orders in 30 and especially 29 are evidenced by higher rotation barriers about these bonds: ~11 Kcal/mole in 15, ~18 Kcal/mole in 30, ~21 Kcal/mole in 29 [90].

29   30

The X-ray studies on salts of heteroatom-containing π-carbanions include lithium 3,3-dimethyl-1-buten-2-olate · THF and lithium cyclopent-enolate · THF (31) [90a, 90b], dipotassium squarate · H_2O (32) [91] and its tetrathio analog [92], diammonium croconate (33) [93], several metal enolates of β-diketones (e.g. *34* [94]; also [95, 96]), and benzophenone dianion dilithium · TMEDA · THF (*35*) [97]. These studies generally reveal highly symmetrical highly delocalized π systems with almost all of the negative charge on oxygen or sulfur, and the cations coordinating with oxygen or sulfur only. This coordination is most often with unshared (n) electron pairs on the heteroatoms rather than with the π system; e.g., lithium acetylacetonate (*34*) has the linear polymeric structure depicted, with lithiums alternately approximating tetrahedral and square planar coordination to four oxygens *via* n-electrons. In the simple enolate tetramers *31*, which resemble the alkyllithium tetramers *9*, the fourth coordination position of lithium is occupied by a THF molecule in each case, and the bond lengths are similar to those found in enol ethers [90a]. In dianion *35*, on the other hand, there is considerable negative charge on carbon (there are two negative charges and only one oxygen), and while one Li$^+$ is in the plane n-coordinated with oxygen, the other lies above the π cloud in the location shown, giving a shortest C—Li distance of 2.24 Å. The C—O bond in this substance is 1.41 Å long, as expected from its lowered bond order compared to benzophenone.

31   32   33

34   35

Simple enolates and phenolates may be free ions in highly ionizing solvents, but they are ion paired in other solvents and aggregated, especially in nonpolar solvents [86]. The lithium enolate of isobutyrophenone has been shown to be tetrameric in dioxolane solution, no doubt with structure *31* [98]. In 1,2-dimethoxyethane, there is 15% tetramer and 85% of a dimeric structure, presumably *36*, analogous to the alkylmetal dimers *7*. Added LiCl converts both tetramer *31* and dimer *36* into a structure like *31* but with Cl replacing OR at one corner. Addition of a cryptand with a strong affinity for Li$^+$ gives a monomeric enolate ion with much more negative charge on carbon as evidenced by an upfield shift in the ^{13}C NMR of 8 ppm for the charge-bearing carbon atom. *Potassium* enolates are probably *not* tetrameric [90b].

II. Structures

Alkali metal enolates form equimolar complexes with alkali metal alkoxides [99].

36

It has been deduced from NMR studies that α-lithiosulfoxides (37) have an sp² hybridized charge-bearing carbon atom, suggesting resonance stabilization of the type shown [100].

37

III. Preparations

The major methods for preparing carbanions are shown in reactions (3-1) to (3-5). *Unstabilized* carbanions (e.g., simple alkyl, aryl, and vinyl anions) are usually prepared from the corresponding halide by reaction with a metal (3-1a) or an organometallic reagent (3-1b). *Resonance* and *inductively-stabilized* carbanions are more often made from an appropriate precursor RH, with the proton being abstracted by a metal (3-2a) or a basic metal salt (3-2b). Certain alkenes add metals to give dicarbanions (3-3a), and metal salts to give monocarbanions (3-3b). The metal can sometimes be changed in an organometallic by reaction with the desired metal (3-4a) or one of its salts (3-4b). Carbanion rearrangements (3-5), useful for preparing certain carbanions, are discussed in Chapter VIII and the other methods are described in the present chapter in the order indicated below.

$$RX + 2M \rightarrow RM + MX \tag{3-1a}$$

$$RX + R'M \rightarrow RM + R'X \tag{3-1b}$$

$$RH + M \rightarrow RM + 1/2\,H_2 \tag{3-2a}$$

$$RH + MZ \rightarrow RM + HZ \tag{3-2b}$$

$$\underset{|\quad|}{\overset{|\quad|}{C=C}} + 2M \rightarrow M-\underset{|\quad|}{\overset{|\quad|}{C-C}}-M \tag{3-3a}$$

$$\underset{|\quad|}{\overset{|\quad|}{C=C}} + MZ \rightarrow Z-\underset{|\quad|}{\overset{|\quad|}{C-C}}-M \tag{3-3b}$$

$$RM' + M \rightarrow RM + M' \tag{3-4a}$$

$$RM' + MZ \rightarrow RM + M'Z \tag{3-4b}$$

$$R'M \rightarrow RM \tag{3-5}$$

1. From Alkyl Halides

a. With Metals (3-1a)

Magnesium, lithium, zinc, sodium and potassium metals all readily reduce organic halides to metal salts of carbanions. The first three are the most commonly used, since sodium and potassium generally give more coupling products (Wurtz reaction) than carbanion.

13

III. Preparations

Grignard reagent preparations (3-6) are the best known examples of this general type of reaction (X = Cl, Br, or I). Ethers are usually required as solvents. Diethyl ether is most often employed, but for more sluggish reactions the better coordinating tetrahydrofuran or the much higher boiling di-*n*-butyl ether and glymes are utilized. Reactivity of RX decreases in the order RI > RBr > RCl, and alkyl halides are more reactive than aryl halides.

$$RX + Mg \longrightarrow RMgX \tag{3-6}$$

Grignard reagent preparations are sometimes difficult to get started: often a small amount of iodine, 1,2-dibromoethane, or some other very reactive alkyl halide is added to initiate reaction [101]. Ultrasounds have recently been reported to initiate Grignard reactions, even in wet commercial grade THF [102]. Magnesium metal formed *in situ* from $MgCl_2$ and K ("Rieke" Mg) enables facile low temperature preparation of even aryl Grignard reagents [103].

Intermolecular [104] and intramolecular [105] trapping experiments as well as CIDNP results [106] indicate that Grignard reagents are formed through alkyl radical intermediates.

The "Schlenk" equilibrium (3-7) [107] can be driven to the right by adding dioxane, which precipitates MgX_2 and leaves a solution of dialkyl- or diaryl-magnesium [108].

$$2\ RMgX \rightleftharpoons R_2Mg + MgX_2 \tag{3-7}$$

Attempts to prepare allylic and benzylic Grignard reagents in the usual way leads to Wurtz type coupling products (3-8) [109, 110], but by careful control of reaction conditions allylic and benzylic Grignard reagents have been successfully prepared [111]. Coupling can sometimes be avoided by preparing the Grignard reagent in the presence of the desired electrophile (the "Barbier" reaction (3-9) [112]), or by using "Rieke" magnesium [113].

$$\tag{3-8}$$

$$\tag{3-9}$$

Grignard-like reagents can also be prepared from compounds other than alkyl halides. For example, reaction of dimethyl sulfate and magnesium yields such a reagent (3-10) [114]. Also, at elevated temperatures THF is reported to react with "Rieke" magnesium to give insertion into the carbon oxygen bond (3-11) [115].

$$Me_2SO_4 + Mg \longrightarrow MeMgOSO_3Me \tag{3-10}$$

$$\tag{3-11}$$

Calcium [116], strontium [117] and barium [117] are of much less importance than magnesium. They react similarly (e.g., (3-12)) but give greater amounts of coupling, which is minimized by running the reactions below room temperature.

$$\phi I \ + \ Ca \ \longrightarrow \ \phi CaI \tag{3-12}$$

Lithium reacts readily with alkyl and aryl halides without ether solvents but both alkanes and ethers are commonly employed as solvents [118]. Organolithiums are generally created from the chloride or bromide, as alkyl iodides tend to give more coupling products [119]. Trace impurities of sodium in the lithium are necessary for the formation of organolithium reagents [120], and for the preparation of t-butyllithium, a copper powder coating on the lithium has been recommended [54].

Preparation of allyl- and benzyllithium by this route leads to coupling [110], but allyllithium is readily produced by reduction of allyl phenyl ether with lithium metal (3-13) [121]. Organolithiums can be prepared by analogous reduction of alkyl phenyl sulfides [122, 123]; α-lithioethers have been made in this way using Li naphthalenide (3-14) [124].

$$\text{\Large \diagup}\!\!\diagup\!\!^{O\phi} + Li \ \longrightarrow \ \text{\Large \diagup}\!\!\diagup\ Li^+ \ + \ LiO\phi \tag{3-13}$$

$$RO\diagdown\!\!\diagup^{S\phi} + Li \ \longrightarrow \ RO\diagdown\!\!\diagdown^- \ Li^+ \ + \ LiS\phi \tag{3-14}$$

Amyl- and phenylsodium are made by slow addition of the corresponding chloride to sodium dispersion at low temperatures (to reduce coupling) [125]; phenylpotassium is made similarly in quantitative yield [126]. Sodium and potassium also reduce aryl alkyl ethers to the corresponding alkylalkali [127, 128].

The Reformatsky reaction of an α-haloester with zinc forms an intermediate organozinc reagent which reacts in situ with an electrophile (3-15) [129]. This reaction has recently been shown to proceed analogously starting with α-halocarboxylate salts [130]. The yields in the Reformatsky reaction are generally better with bromides than with chlorides. Allylic halides can be employed [131]. Alkyl- and arylzinc compounds, with reactivities comparable to Grignard reagents, can be synthesized from "Rieke" zinc and the corresponding alkyl or aryl halide [132].

$$RO\!\!-\!\!\overset{O}{\underset{\|}{C}}\!\!-\!\!X + Zn + \text{\Large Y}\!\!=\!\!O \ \longrightarrow \ RO\!\!-\!\!\overset{O}{\underset{\|}{C}}\!\!-\!\!OZnX \tag{3-15}$$

b. With Organometallics (3-1 b)

Metal-halogen interchanges are used primarily to prepare organolithium compounds from alkyl and aryl halides; they often give better yields than reactions of the halides with lithium metal due to less side reactions [133] In general the rates decrease in the series I > Br > Cl ≫ F. The reactions are equilibrium processes favoring the more stable carbanion [134]. They are performed at or below room temperature in hydrocarbon solvents or (more rapidly) in ethers [135]. Concerted mechanisms have been postulated due to the observed second order kinetics and salt effect [134]. CIDNP experiments on the reaction of alkyllithiums with alkyl and aryl halides indicate the presence of radical intermediates [136, 137].

This method is used frequently to prepare vinyllithiums (3-16 to 3-18) and aryllithiums (3-19, 3-20) from alkyllithiums; each of these reactions used *n*- or *t*-butyllithium in THF or Et$_2$O at −70 to −100 °C. Reactions (3-18 to 3-20) show that carbanions can be prepared by this method even in the presence of functional groups which readily react with carbanions at room temperature. In (3-18), when the chloride was used instead of the bromide, the Si—O bond was cleaved.

(3-16) [138, 139]

(3-17) [140]

(3-18) [141]

(3-19) [142, 143]

(3-20) [144]

Many dicarbanions have been made this way as well (3-21, 3-22).

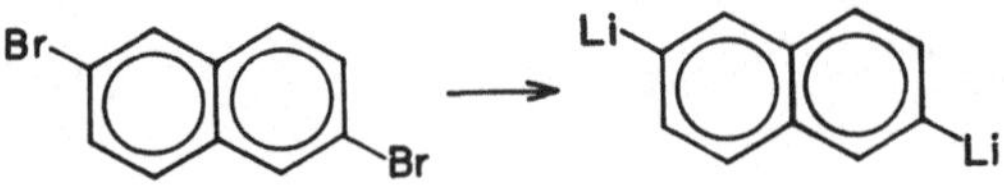

(3-21) [145]

(3-22) [146]

2. From Alkanes by Proton Abstraction (3-2)

a. Acidities of CH Protons

Deprotonation is the method used most often for the preparation of resonance- and inductively-stabilized carbanions; it is thus important to know the acidities of substances containing various types of CH bonds [147]. Several thermodynamic acidity scales have been offered, such as Cram's MSAD scale (McEwen-Streitwieser-Applequist-Dessy) [1], which includes measurements made in various solvents, and Bordwell's scale (Table 2) of values in DMSO. Such scales provide a basis for choosing an appropriate base for a particular CH acid. For example, from the last two entries in Table 2, if kinetics permit, it should be possible to abstract a proton from dimethyl sulfide with an alkyllithium; this has been shown to be the case [162].

Table 2. Bordwell's Acidities of Carbon Acids in DMSO

Acid	pKa	ref	Acid	pKa	ref
$\Phi CH(CN)_2$	4.2	148	ΦSO_2Me	29.0	150
ΦCO_2H	11.0	149	MeOH	29.0	152
$CH_2(CN)_2$	11.1	148	$(\Phi_2P)_2CH_2$	29.9	154
$MeCH(CN)_2$	12.4	150	$MeCO_2Et$	30–31[b]	157
$(MeCO)_2CH_2$	13.3	148	Φ_3CH	30.6	151
			$(\Phi S)_2CH_2$	30.8	154
$\overline{CH_2CH_2SO_2CH_2SO_2CH_2}$			ΦSO_2Et	31.0	150
	15.5	148	Me_2SO_2	31.1	155
$(MeO_2C)_2CH_2$	15.7	149			
$EtNO_2$	16.7	150	$\overline{CH_2CH_2SCH_2SCH_2}$	31.1[a]	158
$MeNO_2$	17.2	150	MeCN	31.3	155
cyclopentadiene	18.0	151	H_2O	31.4	152
ΦOH	18.2	152	Φ_2CH_2	32.1	151
indene	20.1	153	Me_3COH	32.2	152
p-nitrotoluene	20.5	149			
thiourea	21.1	149	$\overset{\displaystyle O}{\underset{\displaystyle \parallel}{\Phi SMe}}$	33[b]	159
ΦCH_2CN	21.9	149			
fluorene	22.6	149	$\overset{\parallel}{NMe}$		
$(\Phi S)_3CH$	22.5	154	Me_2SO	35.1	155
$\Phi COEt$	24.4	150	H_2	36[a, b]	160
$\Phi COMe$	24.7	150	NH_3	41[b]	151
$MeCONH_2$	25.5	150	ΦMe	42[b]	161
cyclohexanone	26.4	149	propene	43[a]	158
Me_2CO	26.5	155	pyrrolidine	44[b]	151
urea	26.9	149	ΦSMe	49[b]	149
$\Phi C\equiv CH$	28.7	156	CH_4	55[b]	151

[a] Solvent other than DMSO
[b] Estimated

Aromaticity can be important in stabilizing carbanions, as illustrated by the very low pKa of cyclopentadiene; its conjugate base, cyclopentadienyl anion, has been estimated to have 24–27 Kcal/mole of aromatic stabilization [151]. Homoaromaticity appears not to be very important in homocyclopentadienyl anions [163, 164], but does help to stabilize homocyclooctatetraene dianions, which form easily and exhibit large ring currents [165, 166]; calculations indicate that it should not be as important in anions as in cations [167].

From Table 2, an ordering of the ability of groups to stabilize a negative charge on an adjacent carbon atom can be obtained: $-NO_2$ > $-(C=O)R$ > $-(C=O)OR$ > $-CN$ ~ $-(SO_2)R$ > $-(SO)R$ > $-\Phi$ > $-SR$. The most strongly stabilizing groups act primarily by resonance; all on this list do except for $-SR$, which stabilizes by polarization [158] and possibly $-(SO_2)R$ and $-(SO)R$ (the latter has an sp^2 hybridized S, which suggests a contribution from resonance stabilization [168]. Ylides (38) [169, 170] are stabilized partly through resonance and partly through inductive effects Certain groups such as $-O(C=O)R$ stabilize partly through inductive effects and partly through chelation (3-23) [171].

$$^-\!\overset{+}{P}R_3 \longleftrightarrow\ =\!\!PR_3$$
$$38$$

$$(3\text{-}23)$$

α-Alkyl groups destabilize simple alkyl, nitrile, and sulfone anions, but stabilize enolates and nitroates; these are complex blends of hyperconjugative steric, and polarizability effects [150] but can be rationalized at least partially on the basis that in enolates and nitroates, the negative charge is very heavily on oxygen rather than carbon, and in enolates and nitroates the methyl group is essentially on a double bond; methyl groups are known to stabilize double bonds.

Almost any type of non-metal atom bonded to a carbon atom will make a proton attached to that carbon acidic enough to be abstracted by very strong bases such as butyllithium to give an essentially σ carbanion. Example in which the metallation is selective enough to be useful include the following groupings attached to the CH: $-SiR_3$ [172, 173], $-NRCH=NR$ [174] $-NR(C=O)OAr$ [175, 176], $-NR(P=O)NR_2$ [177], $-\overset{+}{N}\equiv C^-$ [178, 179] $-PAr_2$ [180], $-\overset{+}{P}R_3$ [169, 170], $-(P=S)Me\Phi$ [181], $-O(C=O)Ar$ [176 182, 183], $-\overset{+}{S}R_2$ [184, 185], $-S(C=S)NR_2$ [186], $-(S=NTs)R$ [187, 188] and $-Se\Phi$ [189, 190].

That two stabilizing groups are better than one is evident from th acidities in Table 2. The many combinations of two such activating grouping which have been successfully employed to make σ carbanions include 2-SiMe$_3$ [191], $-SiMe_3$ and $-OMe$ [192], $-SiMe_3$ and $-S\Phi$ [193], $-SiMe$ and $-Se\Phi$ [194], $-SiMe_3$ and $-Cl$ [195, 196], 2-PΦ_2 [154], $-(P=O)\Phi$ and $-NR_2$ [197], $-(P=O)\Phi_2$ and $-OR$ [198], $-(P=O)\Phi_2$ and $-S\Phi$ [199

$-(P=O)(OEt)_2$ and $-Cl$ [200], 2$-OR$ [201], 2$-SR$ [154, 158, 201, 202], $-SR$ and $-S(O_2)-$ [203], $-S(O)-$ and $-Cl$ [204], and 2$-Cl$ [205].

A few cases of three such stabilizing groups are known: 3$-SiMe_3$ [191], 3$-S\Phi$ [154, 206] and 3$-Cl$ [207].

From the table it is evident that carbanions increase in stability with an increase in the amount of s character at the carbanion center such that $sp > sp^2 > sp^3$. Equilibrium acidity measurements indicate the following expanded order for various types of sp^2 and sp^3 groupings: phenyl > vinyl > cyclopropyl > methyl > ethyl > isopropyl > t-butyl [208]. Cyclopropanes owe their enhanced acidity to the increase in s character in their CH bonds relative to simple alkanes [209].

Heteroatom-containing groupings can control which proton is removed from an alkene (3-24) [171]. Examples are known where Y = $-NR_2$ [210, 211], $-\overset{+}{N}{\equiv}C^-$ [212], $-OR$ [213], $-SR$ [214], and $-Cl$ [171].

$$\text{---}^Y \longrightarrow \overset{Y}{\underset{M}{=}\!\!\!\!\big\langle} \qquad (3\text{-}24)$$

Activation of a CH bond in an aromatic ring by a heteroatom, leading to ortho-metallation, has become increasingly important for the preparation of substituted arenes [171]. If the heteroatom is in the aromatic ring (3-25, Y = NR, O, or S) [215—217] it stabilizes largely through its inductive effect. Two heteroatoms are better than one (3-26) [216], and polymetallation is possible (3-27); furan and thiophene give only the dimetallation product shown, whereas N-methylpyrrole gives also the 2,4-dimetallation product. Pyridines usually undergo addition faster than metallation, so metallated pyridines are made in other ways such as from the corresponding aryl halide by halogen exchange (3-1 b).

$$\text{(3-25)}$$

$$\text{(3-26)}$$

$$\text{(3-27)}$$

Alternatively, the heteroatom can be attached directly to the ring (3-28) [171, 218—222], again acting by induction and perhaps partly through complexation with the metal. Virtually any non-metal atom possessing an unshared pair of electrons (including halogens) will direct the metallation *ortho*. Or, the heteroatom can be further out in the chain, as in (3-29) [171, 218–220, 223–234], in which case complexation with the metal rather than an inductive effect controls the reaction. Among groupings used to activate

19

for and direct metallation to an ortho position on an aromatic ring, amides and sulfonamides are better than methoxy, carboxy, and N,N-dimethylamino [218–220].

$$(3\text{-}28)$$

$$(3\text{-}29)$$

The *rate* of ionization (kinetic acidity) can play an important role in deprotonations since kinetic acidities only roughly parallel thermodynamic acidities. Kinetic acidities measured by a variety of techniques have recently been compiled for many C—H acids [235]. Carbon acids ionize much more slowly than oxygen and nitrogen acids of comparable thermodynamic acidity, presumably because C—H bonds are not so favorably polarized for attack by base as are OH and NH bonds. Among carbon acids, α-protons in nitriles ionized faster than expected, and α-protons in nitro compounds ionized slower.

In the generation of enolates from ketones, weaker bases, aprotic solvents, lower temperatures and excess base favor kinetic deprotonations, while strong bases, protic solvents and excess carbonyl precursor favor the equilibrium products [236]. Generally the more highly substituted enolate is the equilibrium product whereas kinetic control favors deprotonation at the less hindered site, yielding the less substituted enolate [236, 237].

Among alkyllithiums, lesser aggregation is probably responsible for *sec*-BuLi reacting 24 times faster than *n*-BuLi; menthyllithium, which exists in dimeric form, is still faster [12].

Development of high-pressure mass spectrometry, flowing afterglow and ion cyclotron resonance spectroscopy has permitted the measurement of gas phase acidities of C—H acids *without cations or solvents* [238]. Comparison with solution acidities shows the stabilizing effects in carbanions of counterion effects [239] and solvation energies [240] to be on the order of 50–90 Kcal/mole. Rather than as pKa's, gas phase acidities are usually expressed in terms of proton affinity of the anion in Kcal/mole (Table 3). They can be calculated by summing the energy required for *homolytic* dissociation of the acid, the electron affinity of the resulting radical, and the ionization potential of a hydrogen atom (313.6 Kcal/mole). Comparing Tables 2 and 3, some striking differences become apparent. Delocalization plays an extra important role in the gas phase, where toluene is more acidic than water, and fluorene (charge delocalized over more atoms) is more acidic than cyclopentadiene. In the gas phase, *t*-butanol is more acidic than methanol, reversing the trend found in solution. As in solution acidities, the role of alkyl substituents is complicated, although generally α-alkyl groups stabilize negative charge in the gas phase; apparently alkyl groups can stabilize a charge of either sign by polarizing as needed.

Table 3. Gas Phase Acidities of Carbon Acids

Acid	$\Delta H°$, Kcal/mole	ref	Acid	$\Delta H°$, Kcal/mole	ref
CF_3CO_2H	322.7	241	Me_2CO	370.0	241
HCl	333.6	241	$MeCO_2Me$	371.0	242
$CH_2(CN)_2$	336.0	241	HF	371.5	242
ΦCO_2H	338.8	241	Me_3COH	373.3	242
$(MeCO)_2CH_2$	343.7	241	$MeCONMe_2$	373.5	242
succinimide	345.4	241	MeCN	373.5	241
$(EtO_2C)_2CH_2$	348.3	241	Me_2SO	374.6	241
$MeCO_2H$	348.5	241	$HC\equiv CH$	375.4	242
ΦOH	349.8	241	CF_3H	375.6	242
ΦCH_2CN	351.9	241	EtOH	376.1	242
H_2S	352.0	241	Φ-i-Pr	377.5	242
p-nitrotoluene	353.1	241	ΦEt	378.3	242
fluorene	353.3	241	ΦMe	379.0	242
cyclopentadiene	355.5	241	MeOH	379.2	242
Me_2CHNO_2	356.6	241	propene	390.8	243
$EtNO_2$	357.3	241	H_2O	390.8	244
$MeNO_2$	357.6	241	benzene	397.0	244
Φ_2CH_2	364.5	242	NH_3	399.6	244
Me_2SO_2	366.4	241	H_2	400.6	244
MeCHO	367.0	241	CH_4	416.6	244

Since most of the calculations which have been done on carbanion stabilities do not include solvent and cation effects, they generally agree better with gas phase acidity measurements than with solution phase acidities [245–248]; these calculations include many delocalized dicarbanions [248].

b. Base-solvent Systems

That a great variety of base-solvent systems have been used to abstract protons in carbanion-generating reactions is not surprising in view of the tremendous range of acidities of C—H compounds. The range of pKa's of such compounds is even larger than the 4 to 55 indicated in Table 2, for $CH(NO_2)_3$ is more acidic than any compound in Table 2, and other alkanes are somewhat less acidic than methane. As can be seen from Table 2, hydroxide and t-butoxide are strong enough in DMSO to abstract the most acidic proton from most of the compounds in Table 2. Stronger bases like alkyllithiums are required to abstract an allylic proton from propene or a benzylic proton from toluene to produce an allyl or benzyl anion in *high* concentration, but weaker bases like KO-*t*-Bu will pull off such protons to provide *low* concentrations of such anions for double bond isomerizations or deuterium exchange reactions.

The *counterion* makes some difference, as indicated by a factor of 10^3 in the rates of alkene isomerization in DMSO with KO-*t*-Bu being faster than LiO-*t*-Bu; the latter is probably more covalent [249].

The *solvent* can have an enormous effect on the reaction, especially when the comparison is between a hydroxylic solvent and one of the non-hydroxylic solvents (e.g., DMSO) which increase basicity by efficiently solvating cations but not anions. Thus KO-*t*-Bu in DMSO is 10^{14} more basic alcohol-free than in the presence of 5% of *t*-BuOH [250]! Such solvents, with their rate factors for a KO-*t*-Bu catalyzed alkene isomerization relative to diglyme include DMSO (1580), hexamethylphosphoramide (HMPA, 465) N-methyl-pyrrolidone (NMP, 300), and dimethylformamide (DMF, rate about the same as NMP) [249]. Thus DMSO is kinetically fast, but its relatively high thermodynamic acidity renders it unsuitable for very strong bases such as metal amides; HMPA is the solvent of choice for these bases. Among the less active solvents which have been used are alcohols, water, ethers, and hydrocarbons.

Base-solvent systems will now be illustrated, starting with some relatively weak bases and finishing with the most powerful ones.

Classical base-catalyzed aldol condensations often employ hydroxide in water or alcohols to generate low concentrations of enolate ions [251] Quaternary ammonium hydroxides have recently found use as bases in phase transfer systems [252, 253], permitting such reactions as Stobbe (3-30) [254 and Darzens condensations (3-31) [255] under simple, convenient conditions

$$\text{ArCHO} + \quad \overset{CO_2Et}{\underset{CO_2Et}{\diagup}} \quad \xrightarrow[OH^-]{\overset{+}{\phi CH_2NMe_2}} \quad Ar \diagup \overset{CO_2Et}{\underset{CO_2H}{}} \qquad (3\text{-}30$$

$$\overset{O}{\diagdown} + \quad \overset{}{\underset{Cl}{\diagup}} SO_2NR_2 \quad \xrightarrow[OH^-]{Bu_4N^+} \quad \overset{O}{\diagdown} SO_2NR_2 \qquad (3\text{-}31$$

Alkoxides are widely used to generate carbanions in high concentration in malonic, cyanoacetic and acetoacetic ester syntheses, and Dieckmann Stobbe and Darzens ester condensations. Especially useful is potassium *t*-butoxide [256]; being tertiary, it is a stronger base than primary and secondary alkoxides, and also is incapable of hydride transfer reactions which can lower yields significantly when these other alkoxides are used Potassium *t*-butoxide in DMSO will readily deprotonate ketones, partially deprotonate esters, and can be used to generate allylic and benzylic carbanions in very low but often useful concentrations.

There are a variety of non-nucleophilic hindered amines capable of generating carbanions in low concentration. 1,8-Diazabicyclo[5.4.0]-7-undecene (DBU) is one of the most popular, being used in the alkylation and acylation of active methylenes (3-32) [257, 258].

$$\text{(3-32)}$$

Lithium, sodium and potassium amides in liquid ammonia are powerful bases capable of generating dicarbanions from β-diketones [259] and higher carbanions from higher ketones (3-33) [260]. Such bases have been used for isomerizing alkynes (3-34) [261], and for deuterium exchange of hydrocarbons [262]. Lithium diisopropylamide (LDA), generated in situ from n-butyllithium and diisopropylamine, is the most common hindered amide used to prepare enolates in high concentration. Lithium 2,2,6,6-tetramethylpiperidide (LiTMP), lithium dicyclohexylamide and lithium hexamethyldisilamide are more sterically hindered and often can be employed when LDA fails [263] The lithium amidine salt of 3,3,6,9,9-pentamethyl-2,10-diazabicyclo[4.4.0]dec-1-ene has proven useful in the preparation of lithium enolates where other hindered amides failed (3-35) [264].

$$\text{(3-33)}$$

$$\text{(3-34)}$$

$$\text{(3-35)}$$

Sodium and potassium hydride are used for generating enolate anions for alkylations and for Dieckmann condensations [265, 266]. Sodium hydride in DMSO produces the dimsyl anion which is a convenient base [267].

The alkali metals themselves are sometimes used for the metallation of very acidic hydrocarbons such as cyclopentadiene and terminal acetylenes [268]; with many compounds such as aldehydes and ketones, reduction (see next section) becomes the main reaction. Sodium and potassium on charcoal, graphite and alumina have recently been employed in the metallation and alkylation of ketones, nitriles and esters [269–271].

Deprotonation of very weakly acidic C—H acids is usually accomplished by metallation with organolithium and organosodium compounds [272–274]. Direct metallation of acidic C—H bonds is often impossible with these organometallics due to their ready addition to groups such as carbonyls.

23

Metallations can be carried out in hydrocarbon solvents, but activating solvents such as ethers or tertiary amines which help peptize the organo-metallic reagent are usually used. For example, *n*-butyllithium/THF allowed preparation of pentadienyllithium from 1,4-pentadiene (3-36); no reaction occurred in hexane [275].

$$\text{(structure)} \longrightarrow \text{(structure)} \qquad (3\text{-}36)$$

Tetramethylethylenediamine (TMEDA) has an even more dramatic effect in activating alkyllithiums and alkylsodiums [276–278], though its "bite" is apparently too small for a potassium ion [33]. *n*-Butyllithium/TMEDA smoothly metallates propene and isobutylene [279], even dimetallating the latter [72, 280]. This base system when given a choice among allyl anions yields predominantly the least substituted one (3-37) [281].

$$\text{(structure)} \longrightarrow \text{(structure)} \qquad (3\text{-}37)$$

It has been demonstrated that the *kinetic* product in metallation of certain substituted olefins and alkylbenzenes with organosodiums and potassiums is metallated at an sp^2 carbon; this rearranges to the thermodynamically more stable (resonance-stabilized) allyl or benzyl anion [282, 283].

Activation of alkyllithiums can also be achieved by addition of alkoxides, especially potassium *t*-butoxide [284–288]. This base system is less nucleophilic than *n*-butyllithium/TMEDA, allowing preparation of carbanions from *conjugated* dienes (3-38) [288]. The much less soluble potassium salt usually precipitates from solution, and the lithium *t*-butoxide may be removed by filtration or decantation. The insolubility of the potassium salt can be an advantage in preserving stereochemistry, as has been done with crotyl anions [289].

$$\text{(structure)} \longrightarrow \text{(structure)}^{2-} \qquad (3\text{-}38)$$

Trimethylsilylmethylpotassium has been found to be a superior metallating agent which is similar in behavior to the potassium *t*-butoxide/*n*-butyllithium base system [287, 290].

These organometallic reagents being very reactive and highly aggregated pose problems in finding appropriate solvents. Diethyl ether, tetrahydro-furan, HMPA, and TMEDA are all commonly used but all do eventually react with organometallics such as *n*-butyllithium [177, 210, 291–293].

Because metallating agents react very readily with air and water, many methods have been devised for the determination of the concentration of the base. The newer methods titrate the organometallic with a diprotic C—H acid which gives a visible end point when one equivalent of base has been added and the highly colored dianion starts to form [294, 295].

3. From Unsaturated Compounds

a. By Reduction (3-3a)

Many types of unsaturated compounds have been two-electron reduced with active metals or electrochemically to form dicarbanions (3-3a). The anion radical from one-electron transfer is a presumed (or in some cases, demonstrated) intermediate. Simple alkenes do not have sufficiently stable dianions to undergo this reaction, but many of their conjugated derivatives do. For example, cyclooctatetraene (antiaromatic) with any of the alkali metals yields cyclooctatetraene dianion (aromatic) [296, 297]. Benzene (aromatic) does *not* yield its antiaromatic dianion, and naphthalene reduces only to its anion radical (its dianion may be prepared by a deprotonation route [22]), but anthracene and most other polycyclic aromatics readily reduce to the corresponding dianions [298]. 1,3-Dienes are reduced by magnesium to butadiene dianions (e.g., 3-39), which have found use in terpene synthesis [299–302]. When 1,3-pentadiene is reduced with alkali metals, the anion radical dimerizes to give a bis-allyl anion which abstracts protons from two 1,3-pentadiene molecules, yielding two pentadienyl anions and a mixture of $C_{10}H_{18}$ dienes (3-40); this reaction provides an economical route to certain pentadienyl anions from 1,3-dienes and an alkali metal [303].

$$(3\text{-}39)$$

$$(3\text{-}40)$$

Electrochemical methods have been used to "reductively acylate" olefin derivatives such as α,β-unsaturated esters [304, 305] and styrenes (e.g., 3-41) [306] via the corresponding anion radicals and perhaps dianions to various acyl and diacyl derivatives. E can be —CHO (from DMF) or —(CO)R (from Ac_2O or RCN).

$$(3\text{-}41)$$

III. Preparations

There are several examples of the reduction of σ bonds as in (3-42) [132] to give stabilized carbanions; such reactions probably occur by cleavage of the carbon-carbon bond in the intermediate anion radical to give anion and radical, followed by reduction of the latter to give a second anion.

$$(\phi_2 CH)_2 \ + \ 2 \ K \ \longrightarrow \ 2 \ \phi_2 CHK \tag{3-42}$$

b. By Addition (3-3b)

The Michael (3-43) [307], Meisenheimer (3-44) [308], and Chichibabin reactions (3-45) [309] are common examples of this method of carbanion generation, which is only successful for stabilized carbanions.

$$\tag{3-43}$$

$$\tag{3-44}$$

$$\tag{3-45}$$

This method has been applied to formation of enaminates (3-46) [310, 312], benzylic anions (3-47) [313], and pentadienyl anions (3-48) [314]. Anionic polymerization proceeds through a propagation step of this type [315].

$$\tag{3-46}$$

$$\tag{3-47}$$

$$\tag{3-48}$$

Enolate anions resulting from the conjugate addition of organocuprates to α,β-unsaturated carbonyls are recognized as valuable intermediates in natural product synthesis [316, 317].

4. From Other Organometallics by Changing the Metal

a. With Metals (3-4 a)

The most common transmetallation reactions involve organotin and mercury compounds, but analogous reactions for silicon and lead are known. Metal-metal exchange is used especially when halide — free carbanions are wanted. Reactions of type (3-4a) are slow and reversible, and M must be more electro-negative than M' for the reaction to proceed [318]. For example, reduction of an organomercurial with an alkali metal [319] can be used to obtain unstabili-zed (3-49 [320]), stabilized (3-50 [321] and 3-51 [322]), and dicarbanions (3-52 [323]).

$$sec\text{-}Bu_2Hg \ + \ Li \ \longrightarrow \ 2 \ sec\text{-}BuLi \ + \ Hg \tag{3-49}$$

$$(Me_3SiCH_2)_2Hg \ + \ K \ \longrightarrow \ 2 \ Me_3SiCH_2K \ + \ Hg \tag{3-50}$$

$$(\phi CH_2)_2Hg \ + \ Li \ \longrightarrow \ 2 \ \phi CH_2Li \ + \ Hg \tag{3-51}$$

(3-52)

This reaction has recently been used in the conversion of primary alkenes to carbanions through mercurated intermediates (3-53) [324, 325]. Alcohols could be used in place of primary amines.

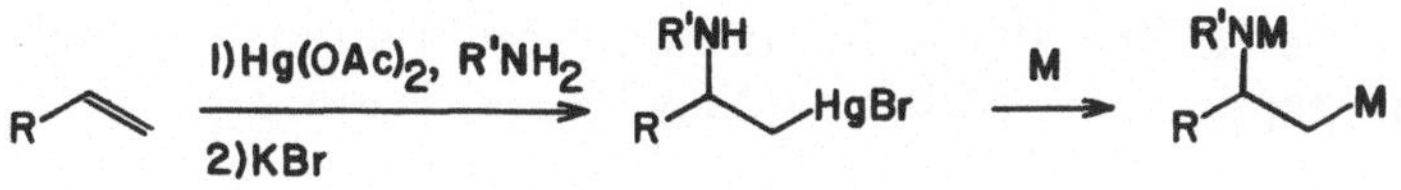

(3-53)

b. With Metal Salts (3-4 b)

Transmetallation of organotin compounds with organolithiums is a useful example of (3-4 b). This reaction is an equilibrium which is driven by formation of the more stable carbanion or by precipitation of one of the products [326]. Vinyllithium [327], many substituted allyllithiums [328–330], and dicarbanions (3-54) [331] have been made by this route.

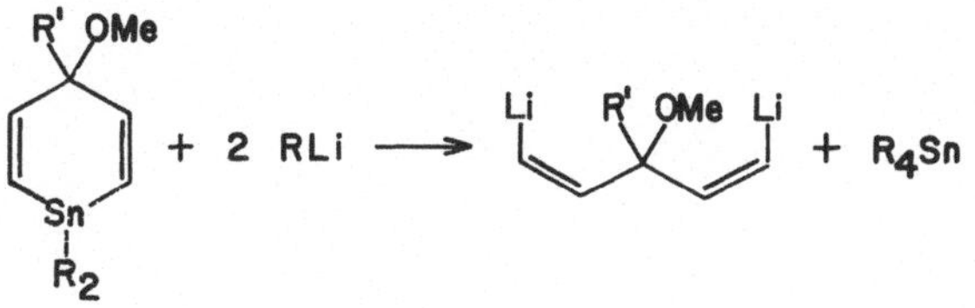

(3-54)

III. Preparations

Alkenyllithiums are conveniently derived from alkenyltin reagents which are readily made by hydrostannation of substituted alkynes (3-55) [332–334].

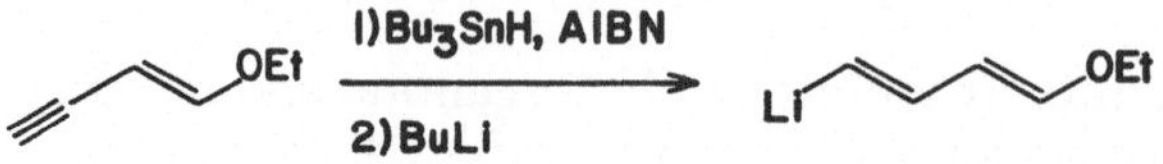

$$
\text{(3-55)}
$$

Alkenyltin compounds are recognized in organic synthesis as latent carbanion equivalents (3-56) [335].

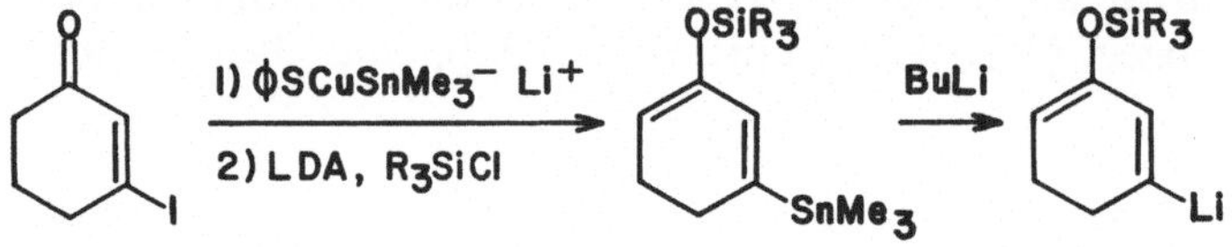

$$
\text{(3-56)}
$$

This same type of reaction is employed for regiospecific generation of enolate anions from the corresponding trimethylsilyl vinyl ether and methyllithium (3-57) [336].

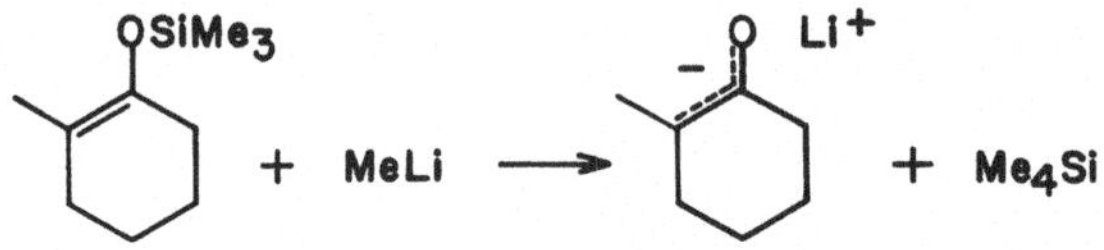

$$
\text{(3-57)}
$$

Allyl anions with quaternary ammonium counterions can be generated and reacted *in situ* with aldehydes and ketones (3-58) [337–339]. This reaction is initiated by the formation of Me₃SiF, and made possible by the great stability of the Si—F bond. The allyl anion does not build up in very high concentration, but is trapped as it adds to aldehydes and ketones.

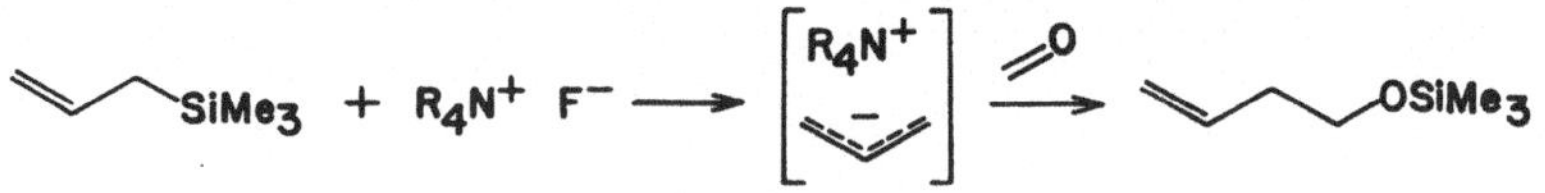

$$
\text{(3-58)}
$$

IV. Reactions of σ Carbanions with Electrophiles

These reactions, summarized in Table 4, have been divided into substitutions (4-1 to 4-21) and additions (4-22 to 4-31). In the former, a leaving group Y usually departs as a bond is formed between R and some non-metallic (4-1 to 4-20) or metallic (4-21) element. Virtually all of the elements except the noble gases can be bonded to carbon in this way. Y^- is most often a halide ion, but many other groupings such as sulfonates, sulfates, and alkoxides have been used. The additions to double and triple bonds form a bond between R and C, N, O, or S, with the resulting anion usually picking up a proton as in (4-22 to 4-24). It should be noted that all of these reactions are limited in scope in that certain other groupings which react readily with carbanions cannot be present, and many have other limitations. In some cases, R′ can be an aromatic group or a hydrogen, and in others, it cannot.

1. Substitution Reactions of Alkyl (sp³) Anions

Abstraction of a proton by an organometallic (reaction 4-1) is an extremely important reaction, as such substances as the butyllithiums are among the most readily available very strong bases and thus are used to generate many other anions, as described in Chapter III. The reaction can also be used to make specifically deuterated hydrocarbons; e.g., the most convenient way to generate CH_3D is to add D_2O (slowly!) in an ether to CH_3Li or CH_3MgI in ether.

Alkylboranes can be produced by (4-2), usually as an intermediate step in converting a carbanion to the corresponding alcohol. Either a dialkoxy-fluoroborane [340] or a trialkoxyborane [341] can be used. Reactions of more complicated carbanions with alkylboranes may be followed by rearrangement of an alkyl group from boron to carbon [342, 343].

One of the most important carbon-carbon bond-forming reactions consists of a carbanion coupling with an alkyl halide (4-3). The use of copper-lithium reagents R_2CuLi, readily prepared from the alkyllithium and cuprous chloride, bromide, or iodide, gives high yields of many unsymmetrical alkanes in cases where the yield directly from the alkyllithium or alkylsodium (Wurtz reaction) is low [344].

A variety of heterocuprates RCuYLi have been used to make more efficient use of R groups; $Y = S\Phi_2$ gives a relatively stable reagent [345]. The mechanism may be simple S_N2, or, especially with very basic carbanions,

Table 4. Reactions of Carbanions R^- with Electrophiles

Substitutions			Additions		
YH[a]	$\xrightarrow{R^-}$ RH	4-1	$\underset{\mid\ \ \mid}{C=CW}$[b] $\xrightarrow{R^-}$ $\underset{\mid\ \ \mid}{RC-CHW}$		4-22
YB(OR')$_2$	$\rightarrow$ RB(OR')$_2$	4-2			
YR'	$\rightarrow$ RR'	4-3	$-C\equiv C-$ $\rightarrow$ $\underset{\mid\ \ \mid}{RC=CH}$		4-23
$\underset{\mid\ \ \mid}{C=C-CY}$	$\rightarrow$ $\underset{\mid\ \ \mid}{RC-C=C}$	4-4	$\underset{\mid\ \ \mid}{C=N}$ $\rightarrow$ $\underset{\mid\ \ \mid}{RC-NH}$		4-24
$-C\equiv C-CY$	$\rightarrow$ $\underset{\mid\ \ \mid}{RC=C=C}$	4-5	$\underset{\mid\ \ \mid}{C=N^+}$ $\rightarrow$ $\underset{\mid\ \ \mid}{RC-N}$		4-25
			$-C\equiv N$ $\rightarrow$ $\underset{\mid}{RC=NH}$		4-26
$\underset{O}{>C-C<}$	$\rightarrow$ $\underset{\mid\ \ \mid}{RC-COH}$	4-6	$\underset{\mid}{C=O}$ $\rightarrow$ $\underset{\mid}{RC-OH}$		4-27
$\underset{\mid\ \ \mid}{YC=C}$	$\rightarrow$ $\underset{\mid\ \ \mid}{RC=C}$	4-7	$O=C=O$ $\rightarrow$ RCO_2H		4-28
			$C\equiv O$ $\rightarrow$ $R_2C=O$		4-29
$\underset{\mid}{C=C=CY}$	$\rightarrow$ $\underset{\mid}{RC-C\equiv C-}$	4-8	$\underset{\mid}{S=CSR'}$ $\rightarrow$ $\underset{\mid}{RSCHSR'}$		4-30
YAr	$\rightarrow$ RAr	4-9	$S=C=S$ $\rightarrow$ RCS_2H		4-31
$\underset{\mid}{YC=O}$	$\rightarrow$ $\underset{\mid}{RC=O}$	4-10			
$YC\equiv N$	$\rightarrow RC\equiv N$	4-11			
YSiR'$_3$	$\rightarrow$ RSiR'$_3$	4-12			
YNR'$_2$	$\rightarrow$ RNR'$_2$	4-13			
YNO$_2$	$\rightarrow$ RNO$_2$	4-14			
YPR'$_2$	$\rightarrow$ RPR'$_2$	4-15			
O$_2$	$\rightarrow$ (see Chapter VIII)	4-16			
S$_8$	$\rightarrow$ RS$^-$	4-17			
YSR'	$\rightarrow$ RSR'	4-18			
YSO$_2$R'	$\rightarrow$ RSO$_2$R'	4-19			
YX	$\rightarrow$ RX	4-20			
YM	$\rightarrow$ RM	4-21			

[a] Y = leaving group, most often halide
[b] W = carbanion-stabilizing group

may involve single electron transfer (SET) from the carbanion to the alkyl halide, giving an alkyl radical and an anion radical which very rapidly decomposes into a halide ion and an alkyl radical; the alkyl radicals can join as desired or give disproportionation or symmetrical coupling products. Bordwell has noted that "the more strongly basic the carbanion, the greater the likelihood that it will react by an electron-transfer pathway rather than an S_N2 pathway" [346]. The reactions of Grignard reagents with alkyl halides are also promoted by the addition of cuprous halides and amines or phosphates [347, 348].

Surprisingly, even tertiary halides like *t*-butyl bromide can be reacted with some carbanions, e.g., as in (4-32), to give coupling products [349a–e]; very likely this sort of reaction goes by an electron transfer mechanism (probably $S_{RN}1$).

$$\text{(structure)} + 2 \text{ (structure)}\!-\!\text{Br} \longrightarrow \text{(structure)} \qquad \text{(4-32) [349]}$$

Heteroatoms are often included in carbanions to guide an alkylation (or other reaction) to a desired position and then removed. Some of the heteroatom-containing groupings are readily replaced by hydrogen (e.g., S by Raney nickel desulfurization [350] or with other reducing agents [351]), hydrolyzed to carbonyl compounds (e.g., dithioacetals to aldehydes and ketones [202, 352, 353] or 1,1,1-trihalides to acids [207]), or eliminated to give alkenes (e.g., —SeΦ [194, 354] and —SΦ [193] by oxidizing and warming, or —S(C=S)NR$_2$ by reacting with methyl iodide followed by LiF/Li$_2$CO$_3$ [186]).

Some S_N2' type reactions (4-4) of carbanions with allyls attached to good leaving groups employ as leaving groups halogen [355], N-ethyl-4,6-dimethyl-2-oxidopyridinium [356], and 2-oxidopyridine [357]. The first two of these leaving groups gave virtually complete S_N2' rearrangement (γ-alkylation), but the last required heavy α-substitution or the S_N2 product predominated. A further sequence for γ-alkylating allyl alcohols *via* a copper complex has been shown to involve *anti* γ-alkylation [358]; if 2-O⁻-benzothiazole is the leaving group, only the *trans*-1,2-disubstituted alkene is formed [359]. The conversion of a propargyl acetal to an allenyl ether by a Grignard reagent catalyzed by cuprous bromide (4-5) [360] is at least formally related to the above, as is a reaction of type 4-8 involving an organoaluminate and allenyl bromide with a cuprous chloride catalyst [361].

The reaction of alkyl carbanions with epoxides (reaction 4-6) can be considered either as an addition or as an intramolecular substitution. The mechanism is ordinarily intramolecular S_N2, with substitution occurring with inversion of configuration at the less hindered carbon atom [362]. Alkyllithiums or Grignard reagents are usually used, but cuprates react similarly [363]. Episulfides sometimes react analogously, though attack can occur at sulfur [364].

Vinyl halides (reaction 4-7) are notoriously sluggish in nucleophilic displacement reactions, since their structures are not conducive to S_N1 and S_N2 mechanisms, and the addition-elimination mechanisms is slow unless some additional grouping is present to stabilize the intermediate carbanion, as in the case $R_2CuLi + R'IC=CH(C=O)R' \rightarrow RR'C=CH(C=O)R'$ [365]. In the absence of such a grouping, a transition metal catalyst containing copper [366], nickel [367], or palladium [368, 369] may serve to break the vinyl-iodine or vinyl-bromine bond and give a good yield of coupled product from a Grignard reagent or alkyllithium. Such reactions generally go with retention of configuration at the double bond. Other possible mechanisms by which vinyl halides may react include halogen exchange to give a vinyl anion and an alkyl halide, which could then react in an S_N2 fashion to give the same coupling product, and elimination to the alkyne followed by nucleophilic addition to the alkyne.

Aryl halides (reaction 4-9) behave toward organometallics very much like the vinyl halides just discussed. S_N1 and S_N2 are disfavored, but addition-elimination involves a resonance-stabilized pentadienyl anion intermediate *39*, and can go very rapidly if a carbanion-stabilizing group such as nitro is present *ortho* or *para* to the halogen. The elimination-addition mechanism now involves a benzyne intermediate *40*, and the R group may go into the *ortho* positions as well as the position to which the halogen was attached. Again, a transition metal catalyst containing palladium may bring about the desired coupling by breaking the carbon-halogen bond [370]. The reaction of methyllithium with iodobenzene to give toluene probably involves halogen exchange followed by displacement [371].

39   *40*

Acylation of a carbanion (reaction 4-10) can proceed readily by an addition followed by elimination of Y^-, but is complicated by addition of further carbanion to the product (probably a ketone; this is a good route to tertiary alcohols of the type $R_2R'COH$ [372]). This further reaction has been avoided in several ways. When an organolithium reagent RLi reacts with a lithium carboxylate, $R'CO_2Li$, the resulting insoluble salt *41* can be quenched by adding to aqueous acid to produce the ketone $RR'C=O$ [373]. Reaction of a Grignard reagent with an anhydride can also give an insoluble precipitate (*42*) which hydrolyzes to the ketone [374]; closely related is the addition product *43* from the S-2-pyridyl thioate, which does not require recycling of R'COOH recovered in the anhydride reaction [375]. Similarly, the N-methyl-N-2-pyridyl formate gives the aldehyde *via* *44* [376]. In HMPA, Grignard reagents are acylated by esters to give ketones if the enolate of the ketone is formed fast enough [377]. Grignard reagents and organolithiums react with tertiary amides to give ketones [378] and aldehydes (from DMF) [379]. An acid chloride can be used directly for acylation if the Grignard reagent or organolithium is first converted to the less reactive cadmium [380, 381], manganese [382], copper [345], or rhodium [383] reagent. $Me_3SiCH_2^-$ with esters gives ketones if two equivalents of the base are used; the ketone is converted to its enolate faster than it suffers nucleophilic attack [384]. (4-10) is not limited to ketone preparation: Formates can give aldehydes, and carbonates can give esters [181].

41   42   43   44

The reaction of carbanions with TsCN (4-11) has been used to prepare nitriles [385].

Silylation with, for example, chlorotrimethylsilane, often proceeds quantitatively and has been used to characterize carbanions (4-12) [172, 386]. Again, however, SET can occur with strongly basic anions to lower the yield [387].

Tertiary amines not easily made by nucleophilic substitution can be synthesized from alkyllithiums or Grignard reagents by reaction with N,N-dialkyl-O-arenesulfonylhydroxylamines (4-13) [388]. Aromatic nitro compounds have been made from aryllithiums and alkyl nitrates (4-14) [389]. Trialkylphosphines can be made as in 4-15 by reaction with dialkylchlorophosphines [390].

The reactions of carbanions with oxygen (4-16) and other oxidizing agents are discussed in Chapter VII. Carbanions react with elemental sulfur to give mercaptides (4-17) [391], and with disulfides (4-18, Y = R'S— [392]), sulfenyl chlorides (4-18, Y = Cl— [393]), or thiocyanates (4-18, Y = N≡C— [394]) to give mixed sulfides RSR'. With phenyl arenesulfonates, alkyl aryl (or diaryl) sulfones are formed (4-19) [395].

Reaction (4-20) of carbanions with halogen-containing species (X_2 [228, 396], TsCl [397], NXS [398], CH_2X_2 [399], CX_4 [400], or C_2X_6 [399]) is sometimes used for replacing an acidic hydrogen by halogen; often, however, the halide RX is more readily prepared than the carbanion R⁻ and it is the reverse reaction (3-1) which is practical. Oxidative coupling (Chapter VII) competes with halogenation in many cases and may predominate.

Reaction (4-21) can be used to attach alkyl groups to any metal, since the most active (alkali) metal can initially be associated with the carbanion, and the reaction favors the organometallic containing the less active metal [401]. This sort of reaction can be used to take advantage of the varied reactivity of carbanions coordinated to transition metals [2]; e.g., (4-33) is a useful synthetic procedure [201, 402].

$$\text{RMgX} \xrightarrow[\text{2) R'I}]{\text{1) Fe(CO)}_5} \text{R(CO)R'} \tag{4-33}$$

2. Addition Reactions of Alkyl (sp^3) Anions

The addition of alkyl anions to alkenes (reaction 4-22) is greatly facilitated by the presence of a carbanion-stabilizing group (W) on the double bond, though one is not essential as shown by the anionic polymerization of ethylene catalyzed by alkyllithiums [403]. Other simple alkenes without carbanion-stabilizing groups fail in this reaction, however, presumably because they lack ethylene's unique combination of an unhindered carbon for attack and a primary carbon at the other end for a relatively stable carbanion. Many alkenes containing carbanion-stabilizing groups, e.g. $CH_2=CHCN$, can be anionically polymerized by catalytic amounts of bases; if the 1:1 adduct is desired, the alkene is added slowly to the alkyllithium.

The carbanion-stabilizing group W can stabilize through induction, resonance, or a combination. Examples of the former are $\Phi Si—$, $\Phi S—$ [404], $\Phi Se—$, and $\Phi_2 As—$ [405]. For resonance stabilization, even a second carbon-carbon double bond (conjugated with the first) will aid greatly, by giving an allyl carbanion intermediate; thus butadiene is readily anionically polymerized [403] or telomerized [406] by organometallics. Some other synthetically useful additions to various conjugated diene systems are shown in reactions (4-34) [407], (4-35) [408], and (4-36) [409].

$$(4\text{-}34)$$

$$(4\text{-}35)$$

$$(4\text{-}36)$$

One of the most important types of carbanion-stabilizing groups W is a carbonyl group, which gives a (resonance-stabilized) enolate ion intermediate. While 1,2-addition of alkyllithiums or Grignard reagents to α,β-unsaturated carbonyl compounds usually predominates over 1,4-addition, when lithium dialkylcuprates (or sometimes simply catalytic amounts of a cuprous halide) are used, 1,4-addition (4-24) is usually favored [410–412] Lithiodithioacetals add 1,2 kinetically and 1,4 thermodynamically [413, 414] A mixed cuprate such as LiCu(R)SΦ, prepared from copper(I) thiophenoxide and an alkyllithium [365], or addition of a trialkylphosphine [412], saves alkyl groups when desired [345]. Of the methyl-copper species, $Me_5Cu_3Li_2$ was found most selective for 1,4-addition [415]. Ways of favoring 1,4-addition

2. Addition Reactions of Alkyl (sp³) Anions

without employing copper are to use zinc [416] or to use an α,β-unsaturated carbonyl compound to which 1,2-addition is sterically difficult, e.g., *45* or *46* [417].

45   *46*

The many other groupings W which promote addition of metal alkyls to alkenes through resonance include $C{=}N-$ [418, 419], $-C{\equiv}N$ ("cyanoethylation" if acrylonitrile itself is used (4-37)) [420], and, in the presence of copper salts and with accompanying reduction to $-NH_2$, $-NO_2$ [421].

$$(4\text{-}37)$$

Metal alkyls add to acetylenes (4-23) somewhat more readily than they do to alkenes since the intermediate carbanion has its charge on an sp^2 hybridized carbon atom; carbanion-stabilizing groups are not necessary, but make the addition faster [54, 398, 422]. Even terminal alkynes may be used with cuprous catalysts, with the carbanion adding *syn* to the non-terminal carbon to give an intermediate vinyl copper reagent *47* which can be reacted with electrophiles E stereospecifically to give trisubstituted alkenes *48* [398]. Though initial additions are *syn*, in some cases the intermediate vinyl anions can be isomerized to provide the other stereoisomer [54]. Alkyllithiums and Grignard reagents react with acetylenic sulfones of type *49* to give acetylenes of type *50*, presumably *via* an addition-elimination mechanism [423].

47   *48*

49   *50*

Many imines are not very stable, but some, e.g. *51*, are sufficiently so that they can be reacted with metal alkyls (4-24) [424–426]. Additions of this type often compete successfully with α-metallation of γ-substituted

35

pyridines, e.g. *52* [231, 427]. Iminium salts, e.g. *53*, can similarly be used to make tertiary amines (4-25, 4-38, 4-39).

51

52

$$R_2CHLi + CH_2\overset{+}{=}NMe_2 \longrightarrow R_2CHCH_2NMe_2 \longrightarrow \longrightarrow R_2C{=}CH_2 \qquad (4\text{-}38)\ [428]$$

53

$$RMgX + (MeO)_2CHNMe_2 \longrightarrow [MeOCH\overset{+}{=}NMe_2] \longrightarrow [MeOCHRNMe_2] \qquad (4\text{-}39)\ [429]$$

Organometallics add to nitriles (4-26) almost as rapidly as to ketones; the imine formed by addition of one mole is not usually stable, but may be hydrolyzed in good yield to the ketone in many cases [430]. In some cases, a second mole of carbanion adds, giving on hydrolysis a tertiary alkyl amine $R_2R'CNH_2$ [430, 431].

The reaction of a carbanion with an aldehyde or ketone to give a 1°, 2°, or 3° alcohol (4-27) is one of the most widely used carbon-carbon bond-forming reactions [432, 433]. Though often thought of as a simple ionic addition, recent evidence indicates that in at least some cases it proceeds by an SET mechanism involving as intermediates R· and the ketone anion radical [434]. Occasional side reactions are enolate ion formation, which usually results in recovered ketone (and can also involve SET [434]), and reduction of the ketone to the corresponding alcohol (4-40); this side reaction may also go by an SET mechanism, at least when a good electron acceptor like a diaryl ketone is used [435]. Considerable progress has been made in the synthesis of optically active 2° alcohols from achiral alkyllithiums and aldehydes using chiral TMEDA analogs [436]. The Barbier reaction, in which an alkyl halide is reacted with Li (much better than Mg) in the presence of an aldehyde or ketone, often proceeds better than the two-step route [102]. There is evidence that in at least some cases the Barbier reaction may not go *via* an organometallic reagent at all, but may follow a radical pathway [437].

$$(4\text{-}40)$$

The additions of heteroatom-containing carbanions to aldehydes and ketones are often followed by useful secondary reactions. The secondary reaction is an olefin-forming elimination in the case of the Wittig (4-41, Y = —PΦ_3^+) [169, 170, 438], Horner-Emmons (4-41, Y = —(P=O)(OR)$_2$) [191, 197, 439–442], Peterson (4-41, Y = —SiMe$_3$) [191, 216, 443–445], and related reactions (4-41, Y = —(C=O)Φ [446], Y = —SnΦ_3 [447], Y = = —S(C=O)OR [448, 449]). These reactions, which have the decided advantage over many other olefin syntheses of giving the double bond in a single known position, probably proceed *via* 4-membered ring intermediates of type *54*. These reactions are versatile enough to permit the synthesis of alkenes possessing —OR [198, 393], —NR$_2$ [197], —SiMe$_3$ [444], —CO$_2$H [441, 442], —CN [393], and —SO$_2$R [446] groupings.

$$\text{—Y} + \text{O=} \longrightarrow \underset{55}{\diagup} \longrightarrow \underset{54}{\square} \longrightarrow = + \text{YO}^- \qquad (4\text{-}41)$$

The addition of anion *55* to an aldehyde followed by pyrolysis leads to a different sort of olefin-forming elimination (4-42) [204].

$$\qquad (4\text{-}42)$$

Certain groupings Y in *55* are γ-eliminated to form epoxides (4-43). Examples are Y = —SMe$_2^+$ [184]; Y = —Cl, in which case an additional activating —SiMe$_3$ group permits acidic hydrolysis to an aldehyde or ketone (4-44) [195, 196]; or Y = —(S=NTs)Φ [187], in which case additional reagent can ring expand the epoxide to an oxetane (4-45) [188].

$$55 \longrightarrow \triangle \qquad (4\text{-}43)$$

$$\overset{\text{SiMe}_3}{\triangle} \longrightarrow \qquad (4\text{-}44)$$

$$\text{=O} \longrightarrow \triangle \longrightarrow \square \qquad (4\text{-}45)$$

The products of addition of α-metallated isocyanides to aldehydes and ketones are readily cyclized to dihydrooxazoles (4-46) [179].

$$\text{—N}\overset{+}{\equiv}\text{C}^- + \text{=O} \longrightarrow \longrightarrow \qquad (4\text{-}46)$$

A standard preparation of carboxylic acids involves reaction of an organo-metallic with CO_2 (4-28) [450]. If a second equivalent of organometallic is present, it may add to the resulting carboxylate anion to give a dianion of type *41*, which will give ketone upon workup [372].

Alkyllithiums react with CO (4-29) to give symmetrical ketones and other products [451, 452].

While carbanions add to the carbon atom of ester carbonyl groups, leading to substitution (4-10) followed by further addition (4-27), the direction of addition to thioester thiocarbonyl groups is reversed (4-30) [352]. Their addition to carbon disulfide (4-31), however, is analogous to their addition to carbon dioxide [453].

3. Vinyl (sp²), Aryl (sp²), and Acetylenic (sp) Anions

These anions undergo most of the reactions described above for alkyl anions (sp³); examples are (4-47) to (4-55). In the reactions of vinyl anions, the configuration about the double bond is retained if the temperature is kept sufficiently low. With vinyl anions as with alkyl anions, the lithium dialkyl-copper reagent is often prepared when the lithium reagent reacts unsatis-factorily, e.g., to get 1,4-additions to α,β-unsaturated carbonyl compounds [454]. Acyl anions are not stable under ordinary conditions [455], though they are probably the first intermediates in the reactions of carbanions with carbon monoxide (4-29) [452]; fortunately many acyl anion equivalents have been devised (Chapter IX).

$$CH_2=CHMgCl + MeCHBrCH_2OMgCl \xrightarrow{CuBr} CH_2=CHCHMeCH_2OMgCl \qquad (4\text{-}47)\ [456]$$

$$R_2C=CHCuR' + NBS \longrightarrow R_2C=CHBr \qquad (4\text{-}48)\ [457]$$

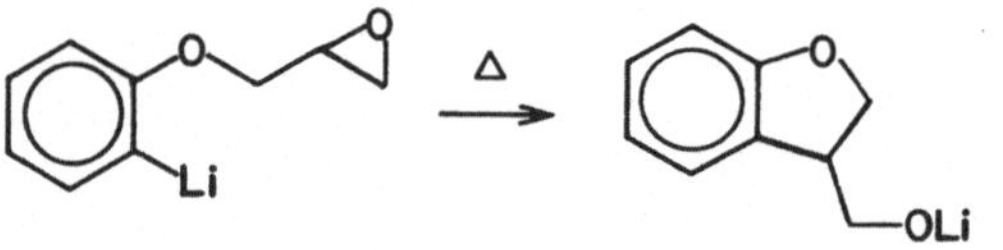

(4-49) [458]

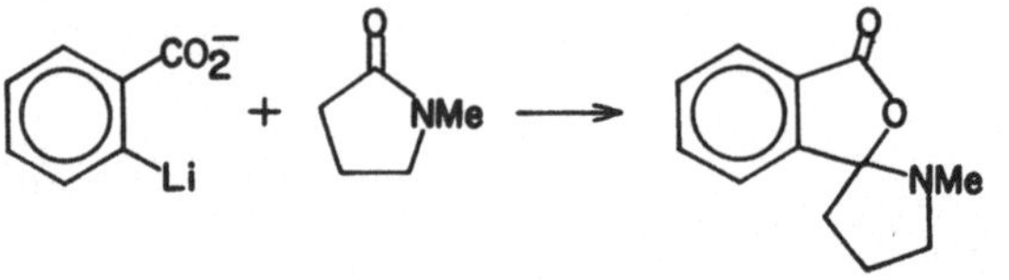

(4-50) [459]

$$R_2C(OAc)C\equiv CLi + R'_3B \longrightarrow R_2C=C=C(R')BR'_2 \xrightarrow{H_2O} R_2CHC\equiv CR' \qquad (4\text{-}51)\ [460]$$

$$RC\equiv CZnCl + ArI \xrightarrow{Pd(P\phi_3)_4} RC\equiv CAr \qquad (4\text{-}52)\ [370]$$

3. Vinyl (sp^2), Aryl (sp^2), and Acetylenic (sp) Anions

$$RC{\equiv}CLi \ + \ NCS \ \longrightarrow \ RC{\equiv}CCl \qquad\qquad (4\text{-}53) \ [461]$$

$$(4\text{-}54) \ [462]$$

$$RC{\equiv}CLi \ + \ R'CO_2Et \ \longrightarrow \ RC{\equiv}C(CO)R' \ + \ (RC{\equiv}C)_2R'COH \qquad (4\text{-}55) \ [463]$$

V. Reactions of π Carbanions with Electrophiles

1. Hydrocarbon π Anions

π type carbanions undergo most of the reactions decribed in the last chapter for alkyl anions; (5-1) to (5-4) are examples with no regiochemical problems

$$CH_2=CHCH_2MgCl \ + \ ClRC=NR' \ \longrightarrow \ (CH_2=CHCH_2)_2RCNHR' \qquad (5\text{-}1) \ [464$$

$$(5\text{-}2) \ [465$$

$$(5\text{-}3) \ [466$$

$$(5\text{-}4) \ [467$$

The main complication which arises with π anions is the site of reaction in unsymmetrical cases; e.g., substituted allyl anion *56* gives *57* by α-attack or *58* by γ-attack. In rate-controlled reactions with electrophiles, attack occurs primarily at the carbon of higher electron density, though steric and other factors no doubt contribute as well. When grouping G stabilizes an α-carbanion (as in *59*) more than it does a double bond (as in *60*), the electron density will be higher at the α-carbon and α-attack to give *57* will predominate this is the case with G = —(C=O)— [468, 469] unless Cu⁺ is the counterion [470], 2-benzothiazole [418], —CN [471], —NO₂ [472], —S— [473], —S(O)– [474], —S(O₂)— [474, 475], and —Cl [476, 477]. When the reverse is true o

40

grouping G, γ-attack to give *58* wins out, as with G = —R (usually) [289], vinyl (usually) [478, 479], —Ar [480], —NR$_2$ [481, 482], and —O— [474].

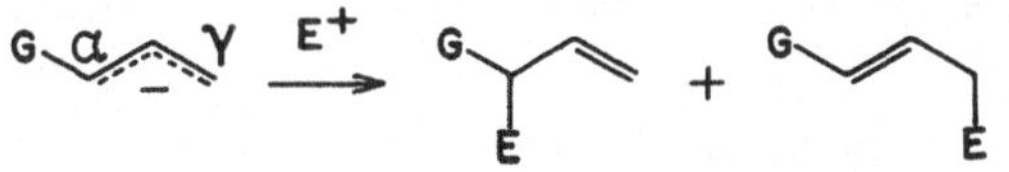

56   **57**   **58**

59   **60**

The reactions of crotyl anion (*56*, G=Me) with electrophiles illustrate the regioselectivity obtained with alkyl groups and in addition the stereochemical results which can be obtained in the γ-alkylation product *58*. Equilibrated crotyllithium (85% *Z*) gives 46% α-butylation and 54% γ-butylation; 85% of the γ product is *Z* [483]. *Z*-Crotylpotassium, prepared at low temperatures from *Z*-2-butene, gives γ-boronation product which is almost 100% *Z*, while *E*-crotylpotassium from *E*-2-butene gives γ-boronation product which is essentially 100% *E* [289].

When prenyl *Grignard* reagents are used, the regiochemistry can be quite well controlled: In the absence of CuI, the γ-alkylation product predominates (92–98%), whereas addition of CuI gives mostly α-alkylation (97–99%) [484]. Interestingly, the α-attack observed when G = —(C=O)— can be converted to γ-attack by adding CuX [485].

The conversion of diene *61* to dienols *62* and *63* with high regio- and stereoselectivity *via* the sequence metallation — B(OMe)$_3$ — H$_2$O$_2$ (5-5 and 5-6) illustrates the reactions of pentadienyl anions with electrophiles [479].

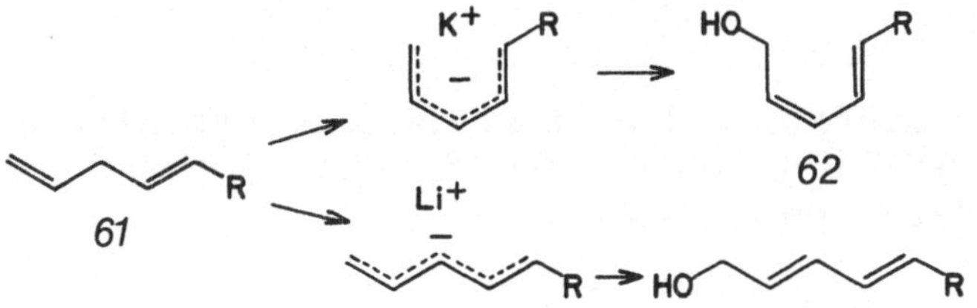

(5-5)

62

61

(5-6)

63

Allylic deprotonation of an alkene followed by protonation of the allyl anion may result in a double bond shift, as in the biomimetric transamination reaction (5-7) [486]. In many cases, the carbanion is not prepared in high concentration, as substances such as alkenes are equilibrated in kinetically strong but thermodynamically weak base-solvent systems such as KO-t-Bu — DMSO (e.g., 5-8) [487].

$$RCH_2NH_2 \xrightarrow{\text{ArCHO}} RCH_2N{=}CHAr \xrightarrow{\text{LDA}} RCH{\overline{\underline{=}}}N{\overline{\underline{=}}}CHAr \xrightarrow{\text{H}^+} RCH{=}NCH_2Ar \tag{5-7}$$

$$\downarrow$$

$$RCHO + ArCH_2NH_2$$

$$\triangleright{-} \longrightarrow \triangleright{=} \tag{5-8}$$

Protonations and other readily reversible reactions can lead eventually to the equilibrium mixture of products, usually different from the rate-controlled mixture. For example, in (5-9), rate-controlled protonation of cycloheptadienyl anion (*64*) gives 1:3 *65:66*, whereas equilibrium control gives 99+ % *66* [314]. The ratio of rate-controlled products depends on the protonating species and conditions; in principle one could find protonating agents and conditions which would give either *65* or *66* exclusively, and in less symmetric cases, one could induce chirality by using a suitable optically active protonating agent. Much less can be done to control the composition of the equilibrium mixture; it will generally vary some with temperature, but the temperature range over which equilibrium can be attained is usually rather restricted by competing reactions. With equilibrium or kinetic control, it may be possible to separate the desired isomer and recycle the undesired one to get more of the desired isomer. In the present case, for example, the unconjugated isomer *65* can be separated from the kinetic mixture and the conjugated isomer *66* recycled by deprotonating it with n-BuLi — KO-t-Bu [288] and again kinetically protonating it.

$$\overset{-}{\underset{64}{\bigcirc}} \longrightarrow \underset{65}{\bigcirc} + \underset{66}{\bigcirc} \tag{5-9}$$

More surprisingly reversible to equilibrium are reactions of prenyl anion with ketones, e.g., (5-10): *67* is formed 4:1 over *68* at short reaction times, but after a long time only *68* is present [488].

$$\underset{}{\bigwedge}{\cdot}{=} + O{=} \longrightarrow \underset{67}{=}{\bigwedge}{-}O^- + \underset{68}{=}{\bigwedge}{-}O^- \tag{5-10}$$

A useful and usually reliable rule for reactions of di- and higher anions is that the first reaction with an electrophile will occur where the last proton came off (assuming the dianion was generated by removal of protons) [259], e.g., (5-11) [489, 490]. The vast majority of examples illustrating this rule involve anions with stabilization from heteroatoms, but the principle applies to hydrocarbon dianions as well. A serious problem with hydrocarbon dianions in this respect is that many of them have very low solubilities (especially dipotassium salts; dilithium salts are more soluble); many cannot be reacted selectively with one electrophile followed by another since the monoanion from initial reaction *is* soluble and reacts faster with electrophile than the dianion.

$$\phi SeCH_2C\equiv CLi \longrightarrow \underset{\underset{Li}{|}}{\phi SeCHC}\equiv CLi \overset{E}{\longrightarrow} \underset{\underset{E}{|}}{\phi SeCHC}\equiv CLi \overset{E'}{\longrightarrow} \underset{\underset{E}{|}}{\phi SeCHC}\equiv CE' \qquad (5\text{-}11)$$

2. Enolate Anions

a. Substitutions

Enolates derived from esters will be included in this section with enolates from aldehydes and ketones.

As with unsymmetrical allyl anions, a question of regiochemistry arises when enolates are reacted with electrophiles; in this case, it is whether reaction will occur at carbon or oxygen. In most cases, the product of bond formation at carbon is desired, and in most cases, this can be achieved by employing conditions which favor it: (1) Li$^+$ rather than K$^+$ or Na$^+$ as the counterion, as lithium enolates tend to associate into dimers and tetramers in which the oxygen is sterically shielded from attack (if it is the monomeric Li enolate which reacts, this also must have a more shielded oxygen than do the other enolates) [86, 98, 491]; (2) A relatively non-polar solvent to favor aggregation and an S$_N$2 mechanism over S$_N$1 (e.g., THF rather than a solvent which strongly solvates cations preferentially like DMF, DMSO, and DME) [98, 492]; (3) Lower temperature to favor aggregation [98, 491]; (4) Avoidance of crown ethers and cryptands, which give relatively free anions [98]; (5) A "softer" electrophile, such as MeI (rather than Me$_2$SO$_4$) or Φ_2MeSiCl (rather than Me$_3$SiCl), which will prefer to react at the "softer" carbon atom rather

$$(5\text{-}12) \ [491]$$

$$(5\text{-}13) \ [493]$$

than at the "harder" oxygen [98, 493]. Reactions (5-12) and (5-13) show how the product can be changed by varying some of these factors.

Polyalkylation of enolates can be a problem if proton exchange between the alkylation product and the enolate occurs very rapidly. This can sometimes be avoided by adding Bu_3SnCl or Et_3Al to generate the tin or aluminium enolate [494] or by adding triethanolamine borate [495]. Enolates of aldehydes can sometimes be alkylated directly, though rapid aldol condensation and polyalkylation can make the yield low [496, 497]. Indirect routes involving alkylation of the enamine [498] or enaminate (Section 5 below) may be preferable. Reactions (5-14) and (5-15) illustrate the alkylation and vinylation of enolates [499].

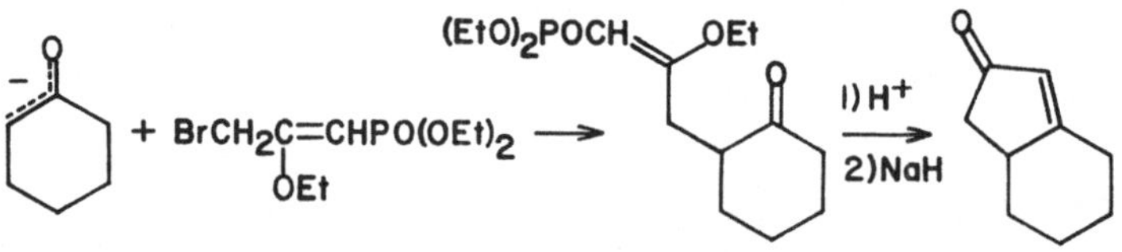

(5-14) [500]

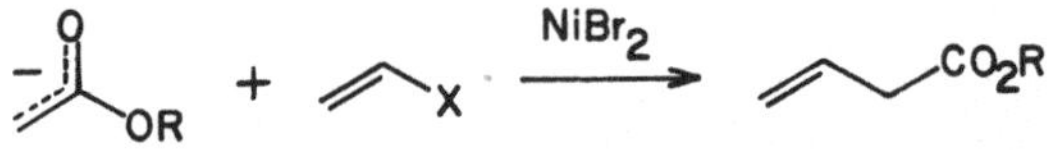

(5-15) [501]

Arylation of enolates, like the vinylation shown in (5-15), does not occur without assistance in breaking the carbon-halogen bond. This may be done with the aid of a transition metal (5-16) or by an $S_{RN}1$ mechanism, in which case the anion radical of the aryl halide dissociates into halide anion and aryl radical (5-17, 5-18) [502–508].

(5-16) [501]

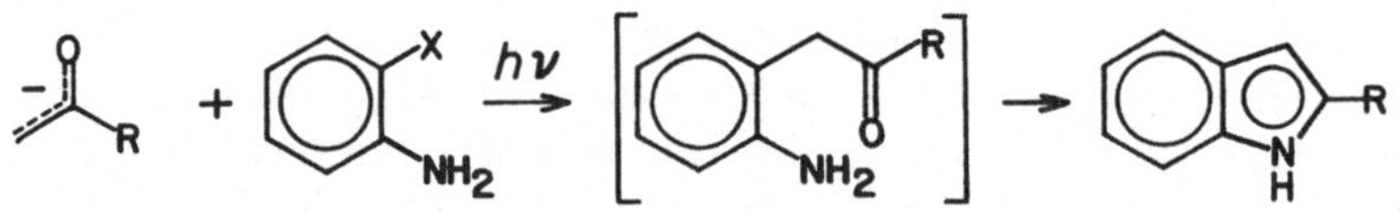

(5-17) [502, 503]

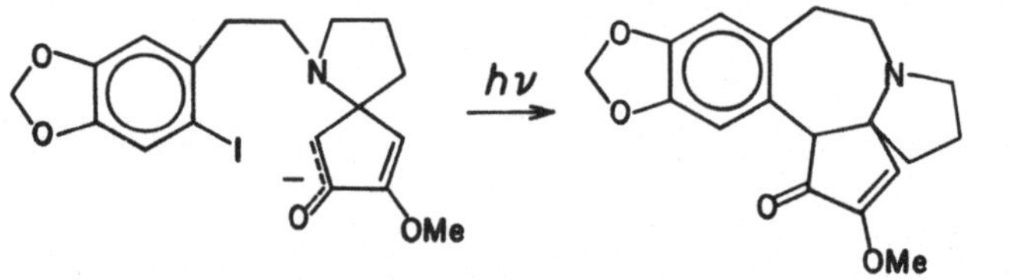

(5-18) [504]

Acylation of an enolate can be accomplished with an acyl chloride (5-19), an acyl cyanide (5-20), an anhydride (5-21), or even an ester (5-22). In the latter case, if the enolate is derived from an ester, the intermolecular reaction is called a Claisen condensation [512] (recently shown to occur even in the gas phase without a counterion [513]) and the intramolecular reaction is a Dieckmann condensation [265]. These two condensations (and a few of the other enolate reactions) are best and most easily run without preliminary generation of the enolate ion in high concentration. The enolate in (5-23) reacts analogously, but O-phosphorylation occurs in (5-24).

$$\text{(5-19) [509]}$$
$$\text{(5-20) [510]}$$

$$\text{(5-21) [511]}$$

$$\text{(5-22) [512]}$$

$$\text{(5-23) [514]}$$

$$\text{(5-24) [515]}$$

b. Additions

The addition of an enolate to an aldehyde or ketone carbonyl group (5-25) is the key step in the base-catalyzed aldol condensation, which is one of the most important carbon-carbon bond-forming reactions [516]. The intermediate β-alkoxylaldehyde or ketone (69) may be protonated, trapped with Me₃SiCl [517], or, under vigorous conditions, caused to eliminate water to give the α,β-unsaturated aldehyde or ketone (70). In cases where mixing the carbonyl components and adding base leads to the wrong aldol product, it may be possible to get the desired product using a preformed enolate [518] or enaminate [519]. This is necessary, for example, if an enolate from an ester is to be added to a ketone containing an α-hydrogen (5-26) [518]. Me₃SiOH may be eliminated instead of water, as in (5-27) [520]. Reaction (5-28) shows the use of an aldol condensation followed by the loss of carboxylate in a vinyl sulfone synthesis [521]. (5-29) shows an aldol followed by γ-elimination of chloride to give α-epoxyester; this is the Darzens condensation [522].

$$\text{(5-25)}$$
$$\text{(5-26)}$$

$$\text{(5-27)}$$

$$\text{(5-28)}$$

$$\text{(5-29)}$$

The stereochemistry of the aldol condensation has been extensively studied recently with a view to developing highly stereoselective aldol methods for synthesizing certain natural products, e.g., macrolide antibiotics [523]. With lithium [524] or boron [525] enolates, the favored kinetic products, obtained at low temperatures, are *threo*-β-hydroxyketone *71* from *E*-enolate *72 via* transition state *73* (5-30) and *erythro*-β-hydroxyketone *74* from *Z*-enolate *507 via* transition state *76* (5-31). The equilibrium product, especially favored by adding ZnCl$_2$, is *threo* (*71*), presumably because R' and R'' are pseudo-equatorial in *77* [526]. With (Et$_2$N)$_3$S$^+$ as the cation, *erythro* product *74* is favored from *either* enolate *72* or *75* (avoiding the necessity of generating a specific enolate), presumably *via* transition states *78* (from *E*-enolate *72*) and *79* (from *Z*-enolate *75*) [517]. For unknown reasons, zirconium enolates give *erythro* products regardless of enolate geometry [527], aluminium *Z*-enolates give considerable *threo* product [528], and either crotyltin stereo-isomer gives *erythro* alcohol *80* which can be converted into *erythro* β-hydroxycarbonyl derivatives *74* [529].

$$\text{(5-30)}$$

72 73 71

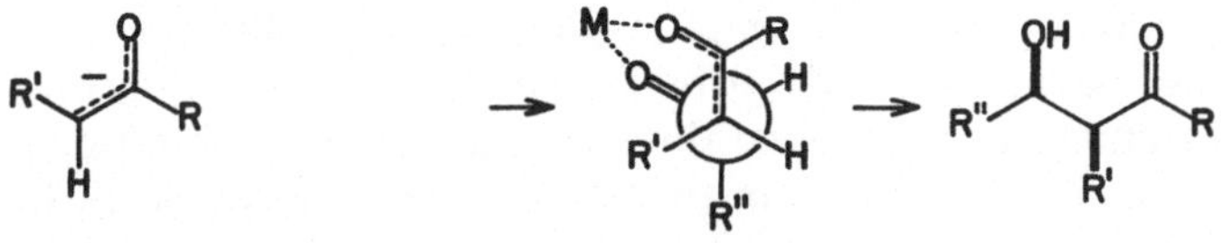

75   76   74   (5-31)

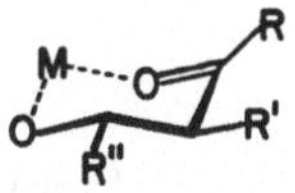

77 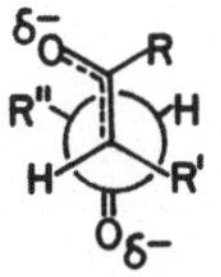   78 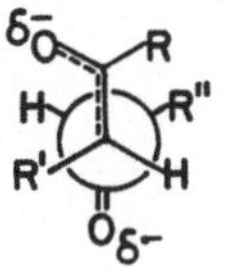   79 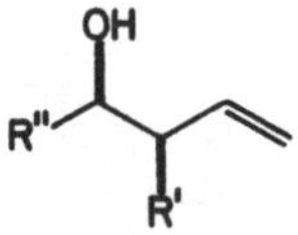   80

Reactions (5-32) to (5-36) show additions of enolate anions to carbon-nitrogen and activated carbon-carbon double bonds. (5-33) is the key step in the base-catalyzed Mannich reaction of secondary amines [535]; in the Mannich the iminium salt is prepared *in situ* by mixing the amine with the aldehyde.

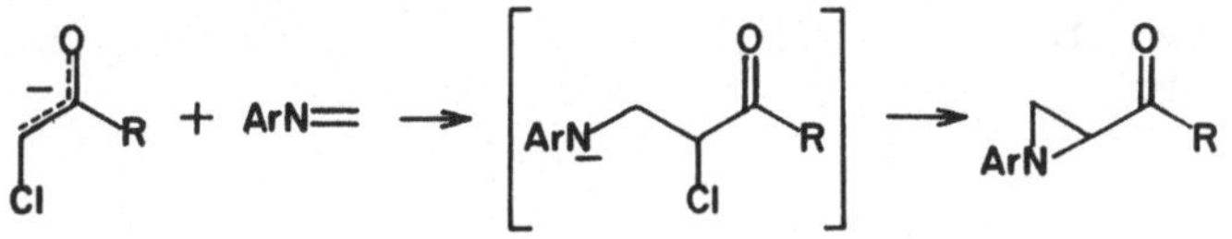

(5-32) [530]

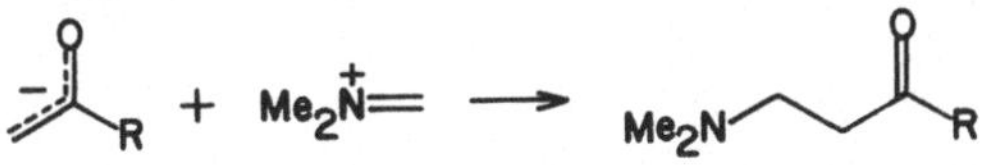

(5-33) [531]

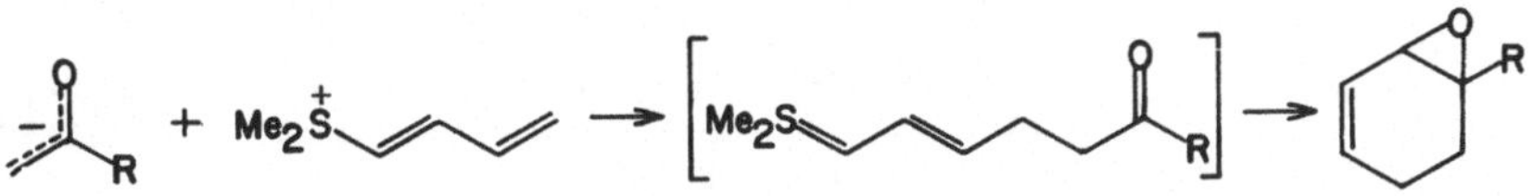

(5-34) [532]

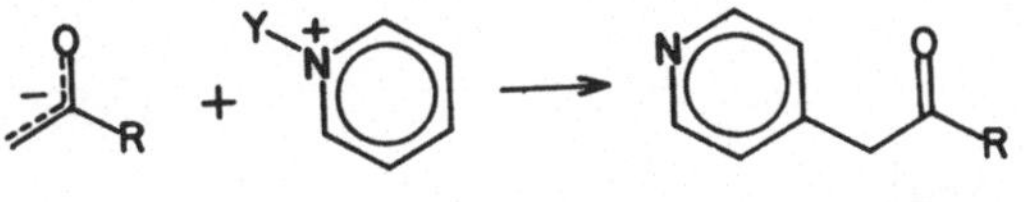

(5-35) [533]

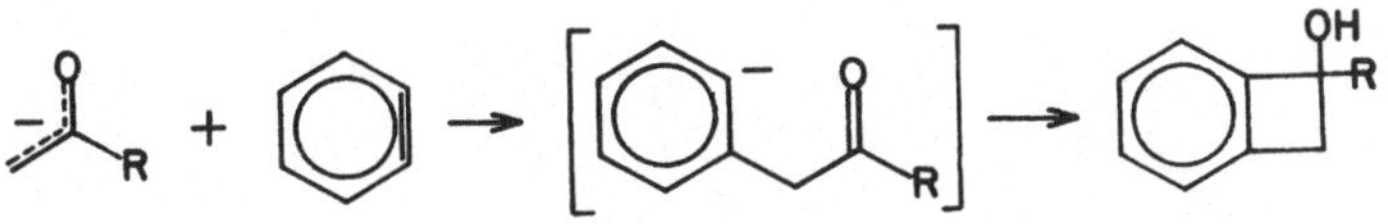

(5-36) [534]

3. Other O-Stabilized Monoanions

The conjugation in enolates can be linearly extended by addition of a carbon-carbon double bond to give dienolate ions *81* and *82*, which usually react with electrophiles under kinetic control in the α-position. In (5-37), the observed γ-attack may indicate thermodynamic control. In (5-38), γ-attack is found when Cu^+ is the counterion, and α-attack when Li^+ is the counterion. If the conjugation is extended in a cross-conjugated manner to give ions *83* and *84*, reaction can occur on oxygen as in (5-39) or in the α-position as in (5-40) and (5-41). Anion *85*, with its conjugation extended by two carbon-carbon double bonds, has been used as an equivalent of 2-metallated-2-cyclohexenone (5-42).

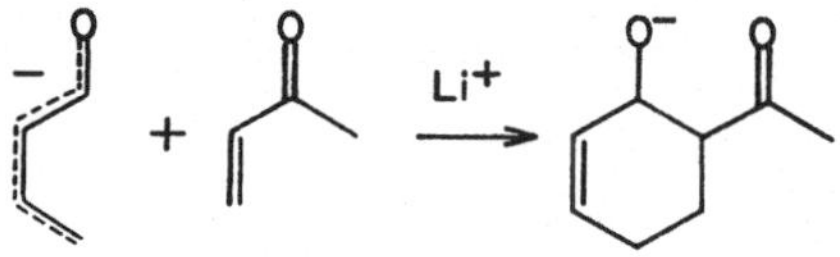

(5-37) [536]

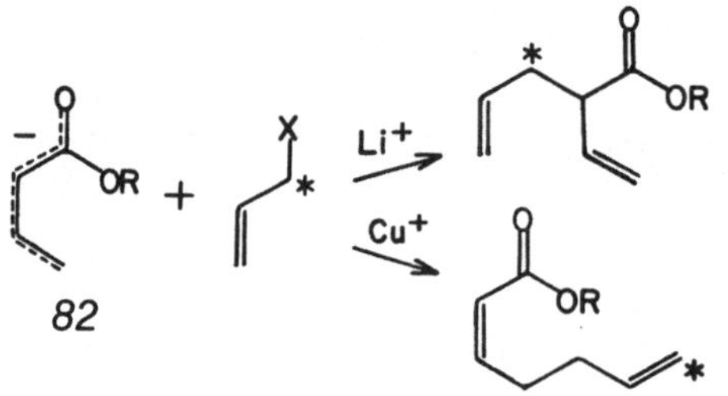

(5-38) [470]

(5-39) [537]

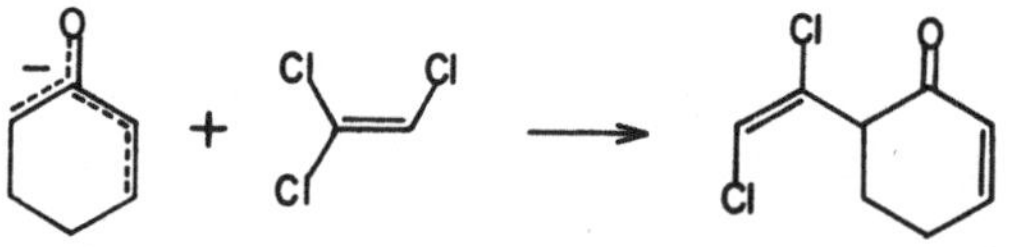

(5-40) [538]

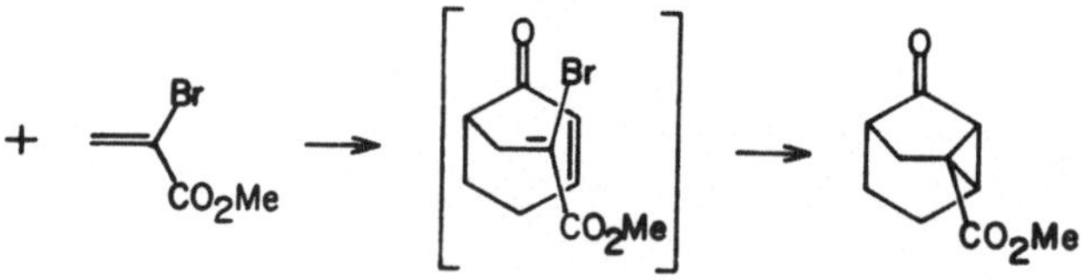

(5-41) [539]

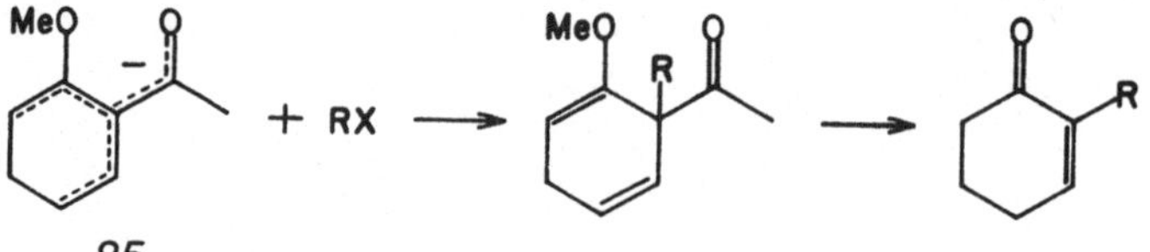

(5-42) [540]

With two stabilizing oxygens arranged as in *86–88*, the anion is relatively easy to prepare and is easy to alkylate on carbon (5-43 to 5-45); (5-44) [542] is the acetoacetic ester synthesis and (5-45) [543] is the malonic ester synthesis.

Some recent variations of these syntheses using di- and trianions will be mentioned in the next section. Two alkene syntheses employing carbanion chemistry of this type are (5-46) and (5-47).

$$\text{(5-43) [541]}$$

$$\text{(5-44) [542]}$$

$$\text{(5-45) [543]}$$

$$\text{(5-46) [544]}$$

$$\text{(5-47) [545]}$$

Two stabilizing oxygens are also present in an anion 89 derived by abstraction of an α proton from a nitroalkane; these anions react with electrophiles on carbon in some cases, e.g. (5-48).

$$\text{(5-48) [546]}$$

Anion *90*, a higher vinylog of *86*, is available commercially with Bu_4N^+ as the counterion for use in synthesis; in at least some of its reactions, e.g. (5-49), initial attack on the α carbon is indicated.

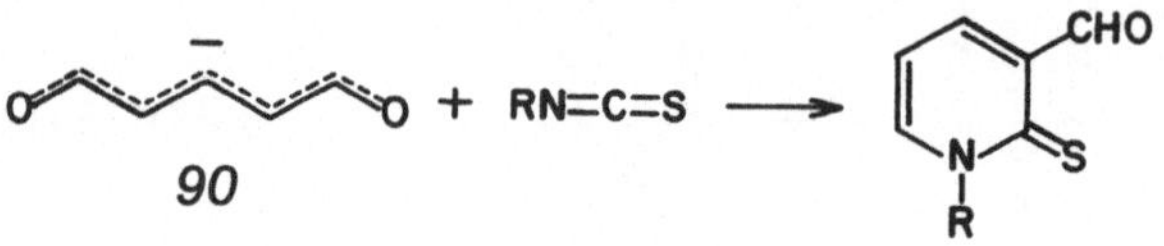

(5-49) [547]

Various types of extended enolate ions have been used as protecting groups (e.g. 5-50) [548–550], with the carbonyl compound being liberated at the end by protonation of the enolate.

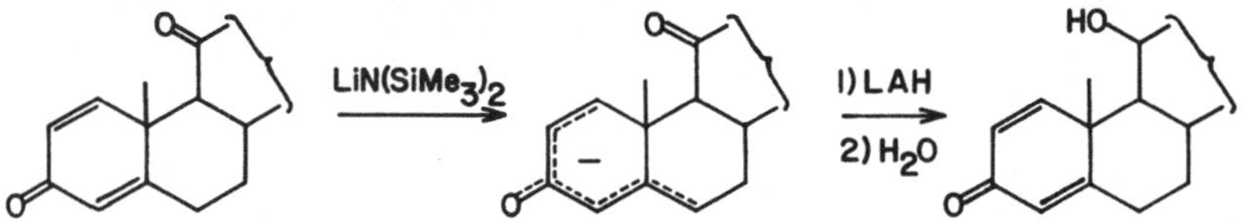

(5-50)

4. O-Stabilized Dianions

Many oxygen-stabilized dianions are synthetically useful, e.g. enediolate dianion *91* in the acetylene synthesis (5-51); initial attack is on oxygen in this case.

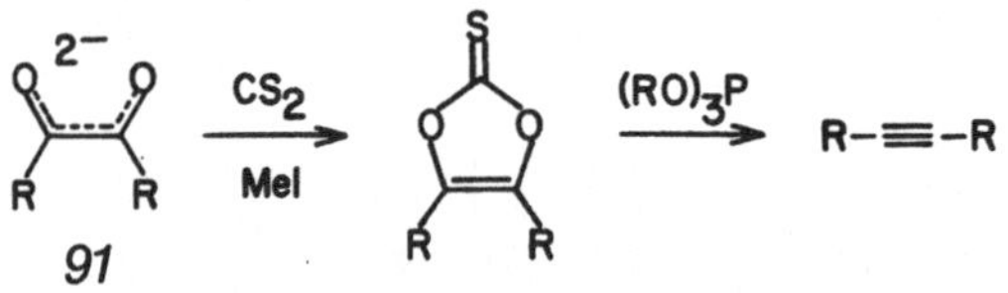

(5-51) [551]

Reactions (5-52) to (5-56) of branched-chain dianions *92*, usually derived by removal of two protons from carboxylic acids, illustrate the usually valid principle that the first site of reaction of a dianion with an electrophile is the atom from which the last proton was removed [259], i.e., at the end carbon in these cases.

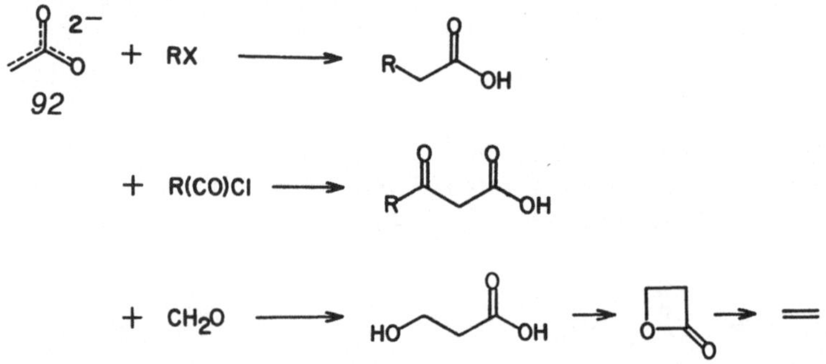

(5-52) [552]

(5-53) [553]

(5-54) [554, 555

$$+ \quad \phi CH{=}CH_2 \quad \longrightarrow \quad \text{[structure]} \qquad (5\text{-}55) \ [556]$$

$$+ \quad CS_2 \quad \longrightarrow \quad \text{[structure]} \quad \xrightarrow[2)\,\Delta]{1)\,MeI} \quad \text{MeS} \quad \text{[structure]} \qquad (5\text{-}56) \ [557]$$

The above dianions *91* and *92* have 4 atoms in a π system bearing the negative charges. Reactions (5-57) to (5-63) of some of the analogous 6-atom systems *93–98* are given below. (5-59) illustrates kinetic *vs.* thermodynamic control in a reversible addition. (5-60) shows that β-diketones and β-ketoesters can be reacted with electrophiles at the γ-position *via* the dianion *96*; this complements well the usual reactions of the monoanions at the α-position. *97* is the dianion (usually with Mg^{2+} as the counterion) of ethyl hydrogen malonate, which is superior to monoanion *88* from diethyl malonate for many "malonic ester" type syntheses because a hydrolysis step is avoided and a monoester is obtained upon decarboxylation.

$$\underset{93}{\text{[structure]}} \quad \xrightarrow[2)\,H^+]{1)\,\triangle} \quad \text{[structure]} \qquad (5\text{-}57) \ [558]$$

$$\underset{94}{\text{[structure]}} \quad + \quad X(CH_2)_nX \quad \longrightarrow \quad (CH_2)_n \underset{CO_2Me}{\overset{CO_2Me}{\text{[structure]}}} \qquad (5\text{-}58) \ [559]$$

$$\underset{95}{\text{[structure]}} \quad + \quad CH_2O \quad \xrightarrow{\text{low temp.}} \text{[structure]} \quad \xrightarrow{\text{high temp.}} \text{[structure]} \qquad (5\text{-}59) \ [560, 561]$$

$$\underset{96}{\text{[structure]}} \quad \xrightarrow[2)\,E']{1)\,E} \quad \text{[structure]} \qquad (5\text{-}60) \ [552, 562\text{-}564]$$

$$\underset{97}{\text{[structure]}} \quad + \quad \text{[structure]} \quad \longrightarrow \quad \text{[structure]} \qquad (5\text{-}61) \ [565]$$

51

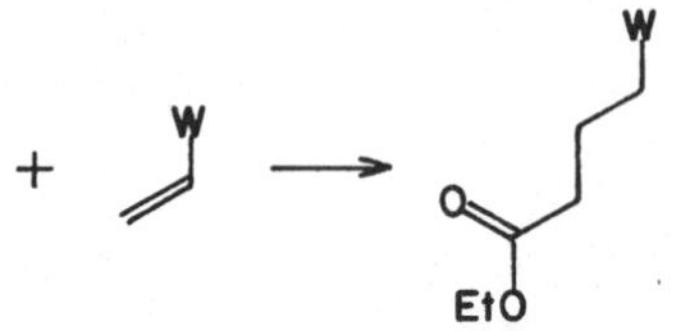

$$\text{+} \quad \overset{W}{\diagup\diagdown} \quad \longrightarrow \quad \text{(structure)} \qquad (5\text{-}62)\ [566]$$

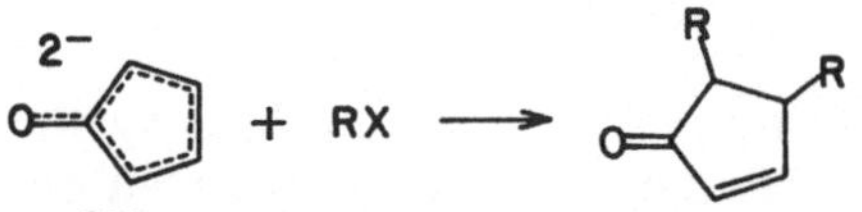

$$\overset{2-}{\text{(structure 98)}} \quad + \quad RX \quad \longrightarrow \quad \text{(structure)} \qquad (5\text{-}63)\ [567]$$

98

Dianions *99* with more extended π systems alkylate as shown (5-64); they have been used as synthetic equivalents of 2-metallated-2-cyclohexenone.

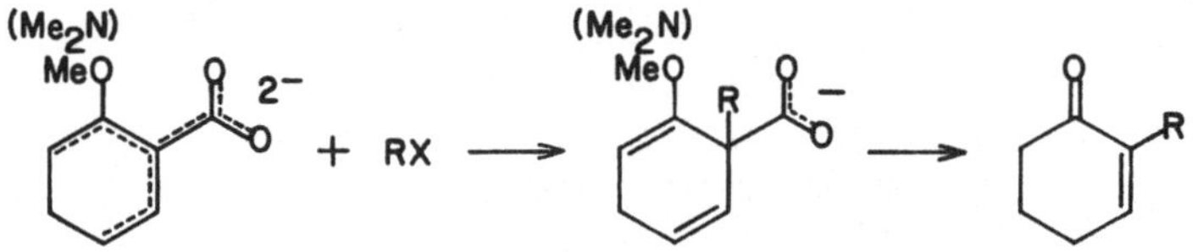

$$\text{(structure 99)} \quad + \quad RX \quad \longrightarrow \quad \text{(structure)} \quad \longrightarrow \quad \text{(structure)} \qquad (5\text{-}64)\ [568,\ 569]$$

99

Dianions *100–102*, with the negative charges in separate systems, have found use in synthesis as shown in (5-65) to (5-67).

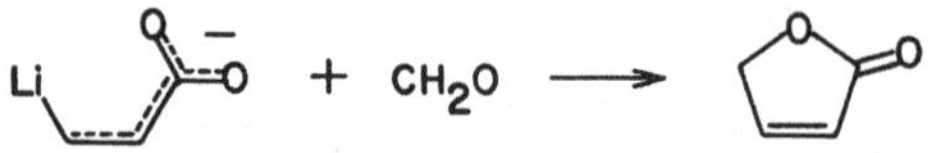

$$\text{(structure 100)} \quad + \quad CH_2O \quad \longrightarrow \quad \text{(structure)} \qquad (5\text{-}65)\ [570]$$

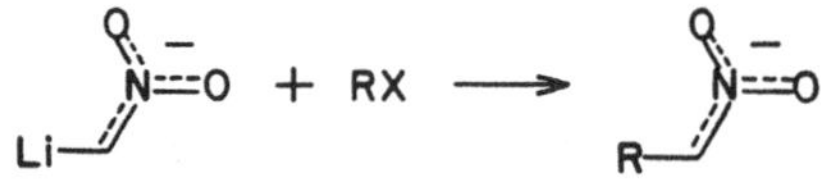

100

$$\text{(structure 101)} \quad + \quad RX \quad \longrightarrow \quad \text{(structure)} \qquad (5\text{-}66)\ [571]$$

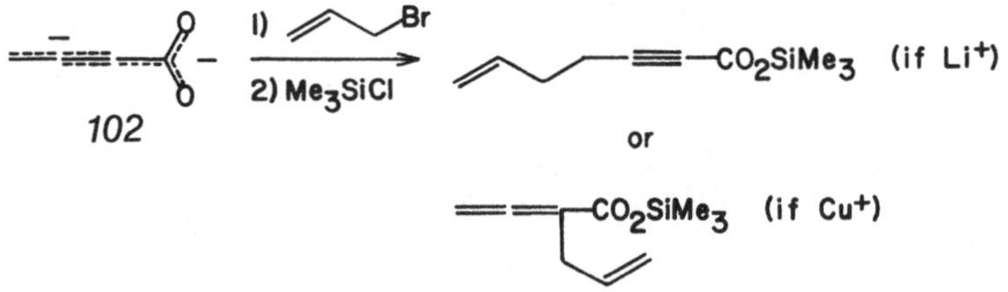

101

$$\text{(structure 102)} \quad \overset{\text{1)} \diagup\diagdown Br}{\underset{\text{2) Me}_3\text{SiCl}}{\longrightarrow}} \quad \diagup\diagdown\equiv-CO_2SiMe_3 \quad (\text{if } Li^+)$$

102 or $\qquad (5\text{-}67)\ [572]$

$$==\text{—}CO_2SiMe_3 \quad (\text{if } Cu^+)$$

5. N-Stabilized Anions

Enaminate ions *103* alkylate readily on carbon as in (5-68) [310–312, 573–579]; some examples of this reaction are given in (5-69) to (5-71). These reactions are most often used when the corresponding enolate reaction is unsatisfactory.

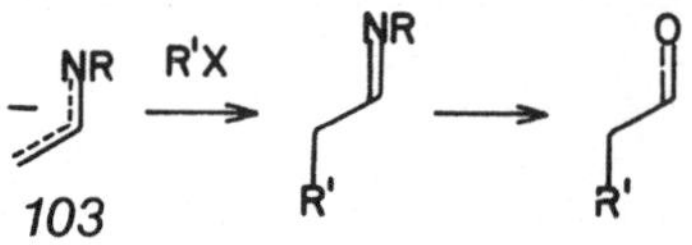

$$\overset{NR}{\text{(structure)}} \quad \overset{R'X}{\longrightarrow} \quad \overset{NR}{\underset{R'}{\text{(structure)}}} \quad \longrightarrow \quad \overset{O}{\underset{R'}{\text{(structure)}}} \qquad (5\text{-}68)$$

103

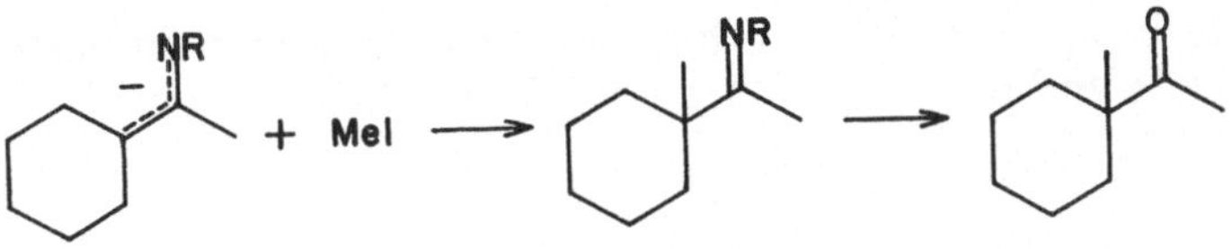

(5-69) [310]

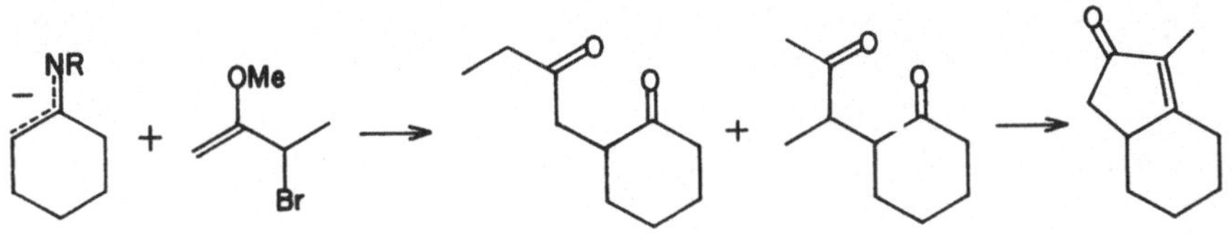

(5-70) [575]

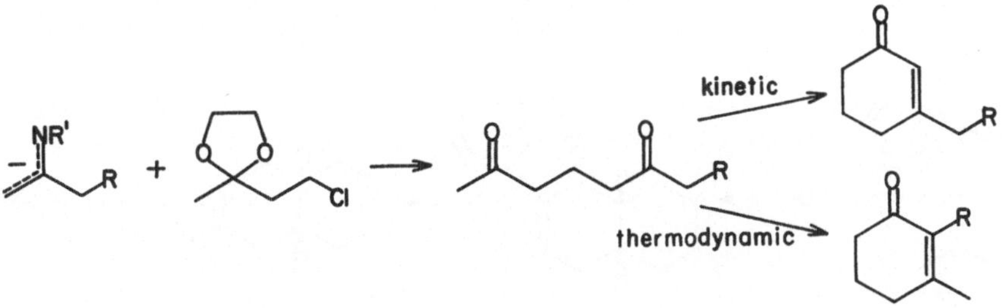

kinetic

thermodynamic

(5-71) [576]

These anions have been used in many asymmetric syntheses of α-alkylated ketones, e.g. (5-72) to (5-74). Many of the groupings on nitrogen contain a heteroatom which can chelate with the metal atom.

* = chiral center

(5-72) [580]

+ R'X ⟶

(5-73) [581]

+ R'X ⟶

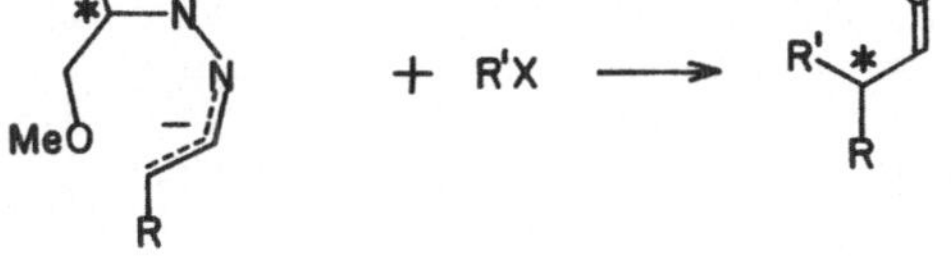

(5-74) [582]

V. Reactions of π Carbanions with Electrophiles

Anions from oxazolines, oxazines, and thiazoles have been used similarly to make carboxylic acids and derivatives, e.g. (5-75) to (5-77).

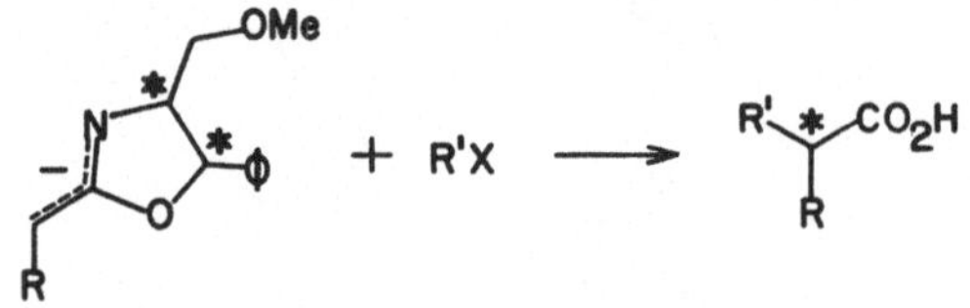

(5-75) [583]

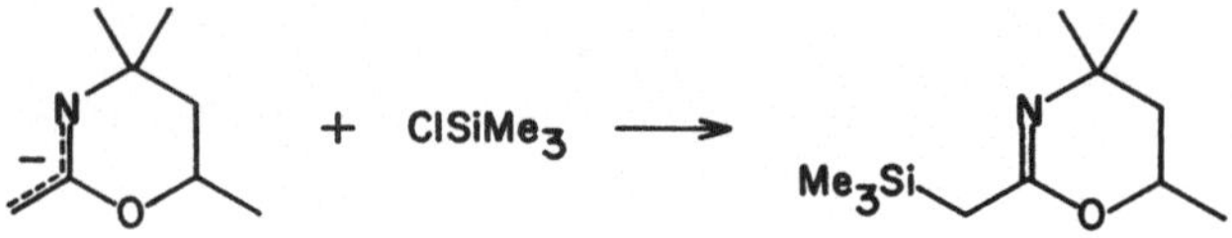

(5-76) [584]

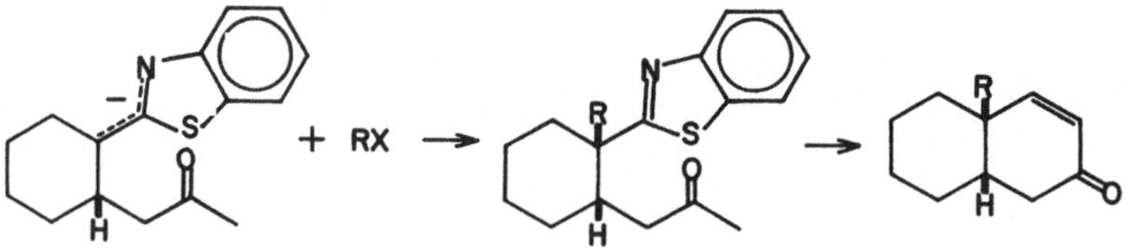

(5-77) [418]

Anions *104* derived from N-nitrosoamines alkylate well on carbon and have been used as equivalents for α-metallated secondary amines (5-78).

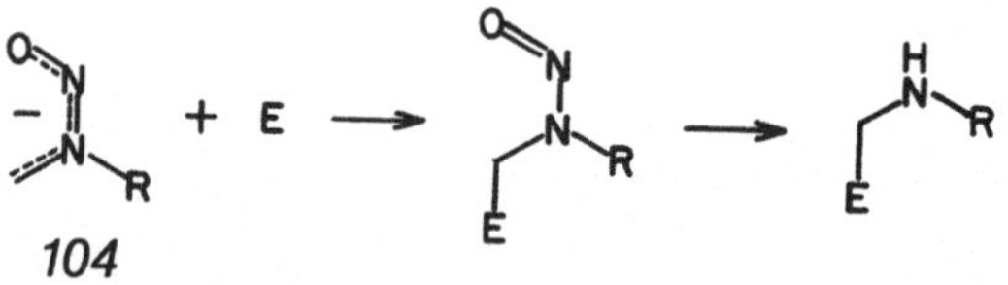

(5-78) [585–589]

α-Metallated acetonitrile (*105*) and some of its derivatives have been used in syntheses as illustrated in (5-79) to (5-81).

$$CH_2\!\!=\!\!\bar{C}\!\!\equiv\!\!N \; + \; (EtO)_2\overset{O}{\overset{\|}{P}}Cl \; \longrightarrow \; (EtO)_2\overset{O}{\overset{\|}{P}}CH_2CN$$

105

(5-79) [590]

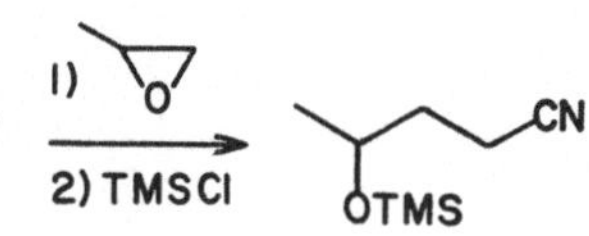

(5-80) [591]

$$R_2NCH\!\!=\!\!\bar{C}\!\!\equiv\!\!N \; + \; E \; \longrightarrow \; R_2N\underset{E}{\overset{|}{C}}HCN \; \longrightarrow \; \underset{E}{\overset{|}{C}}HO$$

(5-81) [592]

The *cis*-dianions *106* derived regiospecifically from *syn*- or *anti*-oximes react with electrophiles on carbon, e.g. (5-82) to (5-84).

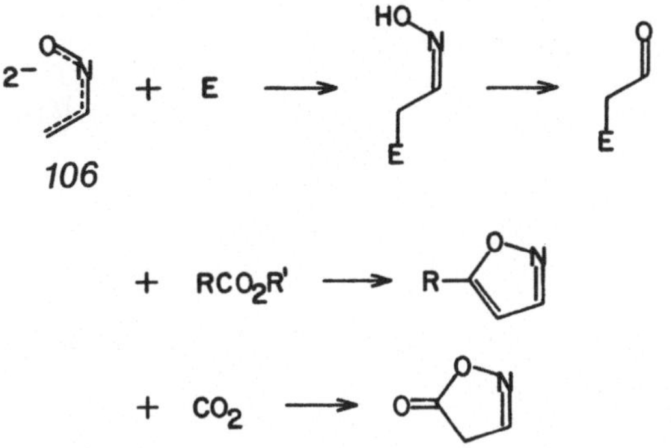

(5-82) [593–595]

(5-83) [596]

(5-84) [597]

An attempted preparation of N-stabilized butadienedianion analog *107* gave instead dianion *108*, stabilized by chelation, as indicated in (5-85).

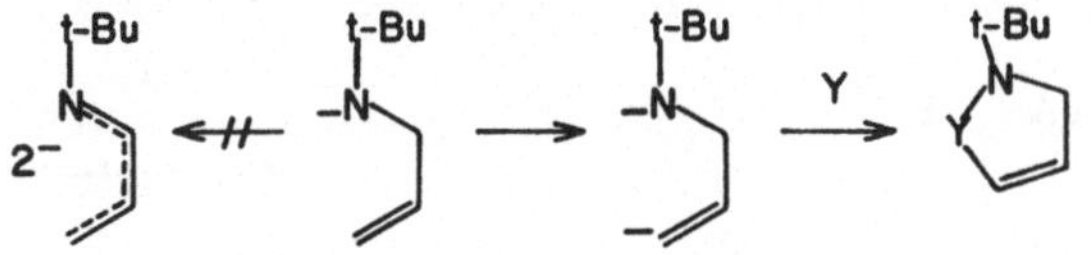

(5-85) [598]

On the other hand, dianions of type *109*, stabilized by two nitrogens, were easily prepared and reacted smoothly with electrophiles at the end carbon atom (5-86 and 5-87).

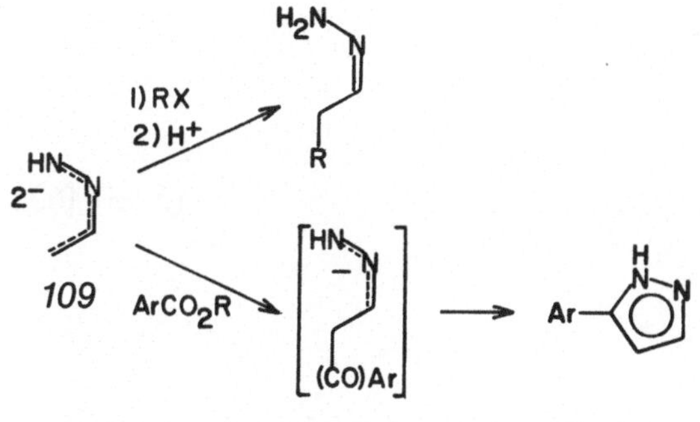

(5-86) [599]

(5-87) [600]

Nitrogen-stabilized hexatrienedianion analogs *110* and *111* react with electrophiles on carbon as shown in (5-88) and (5-89).

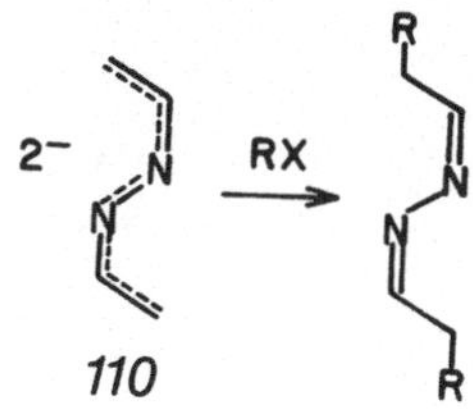

(5-88) [599]

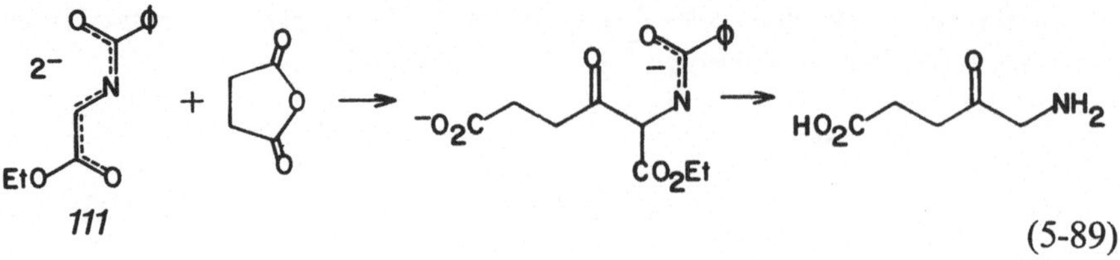

V. Reactions of π Carbanions with Electrophiles

111

(5-89) [601]

6. S-Stabilized Anions

Some analogous S-containing carbanions *112–115* react with electrophiles on carbon as shown in (5-90) to (5-93).

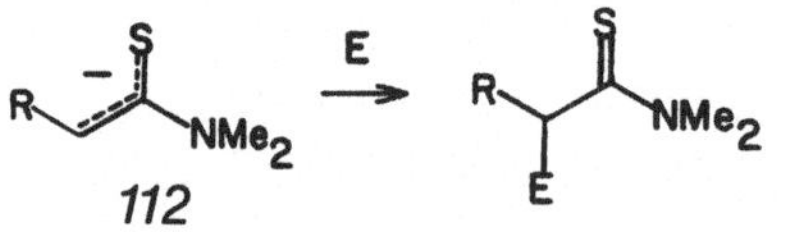

112

(5-90) [602]

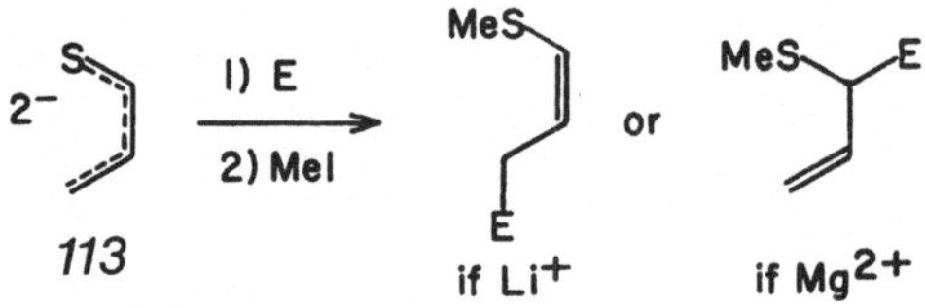

113   if Li⁺   if Mg²⁺

(5-91) [60?]

114

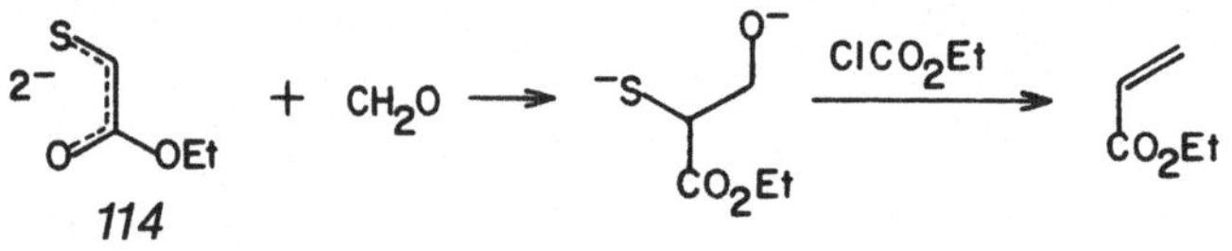

(5-92) [604]

115

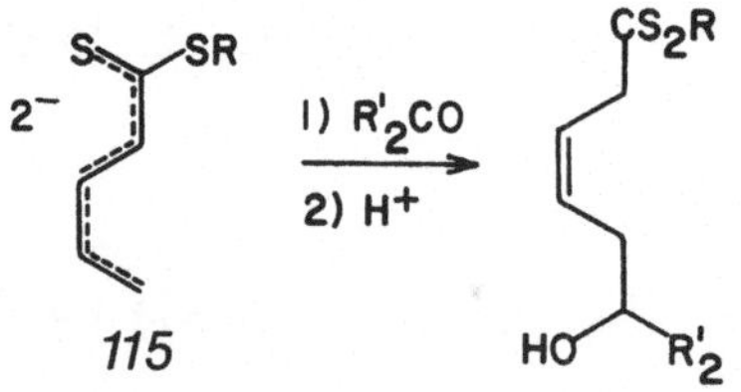

(5-93) [605?]

VI. Eliminations

Eliminations of various leaving groups Y from carbanions are useful routes to carbenes (from α-elimination, (6-1)), alkenes and alkynes (from β-elimination, (6-2a) and (6-2b)) cyclopropanes (from γ-elimination, (6-3)), and vinyl anions (from the Shapiro reaction, (6-4)). Many of these reactions proceed rapidly at room temperature but are slow at dry ice temperatures, thus permitting carbanions containing good leaving groups to be used in substitution and addition reactions at low temperatures [205, 207]. Leaving groups can be halide, RO^-, R_2N^-, or even R^-, H^-, and O^{2-}. Cycloeliminations of carbanions (6-5) are also known.

$$R_2\bar{C}Y \longrightarrow R_2C: \qquad\qquad (6\text{-}1)\ [606]$$

(6-2a)

(6-2b)

(6-3)

(6-4)

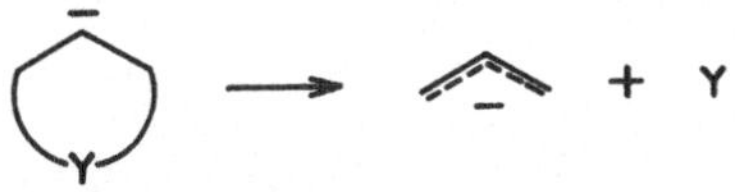

(6-5)

1. α-Eliminations

Examples are (6-6), from the Reimer-Tiemann reaction, and (6-7), which can be avoided at low temperatures. Reaction (6-8) may also go by α-elimination as shown.

VI. Eliminations

$$Cl_3C^- \longrightarrow Cl_2C:$$ (6-6) [606]

$$EtOCH_2^- \longrightarrow CH_2:$$ (6-7) [607]

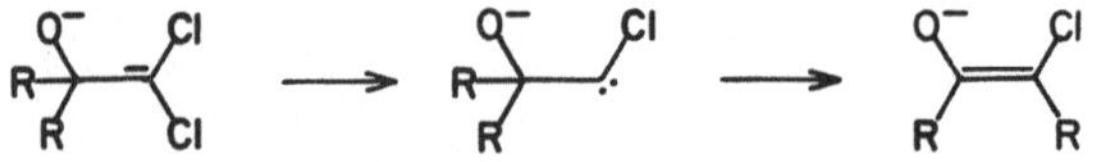

(6-8) [205]

2. β-Eliminations

Reaction (6-2a) is a key step in the formation of alkenes by the ElcB mechanism [608, 609]; e.g., the base-catalyzed aldol condensation (5-25) often involves as a last step the elimination of a β-OH from an enolate ion to give an α,β-unsaturated carbonyl compound (70). This is also a key step in several sequences in which the enolate ion is generated in different ways, e.g., by addition as in (6-9) and by alkylation of a dianion as in (6-10). The former sequence can be carried out with retention of configuration at the double bond [610]. The latter sequence takes advantage of the rare combination of good leaving group ability and α-carbanion stabilizing ability possessed by the CN grouping.

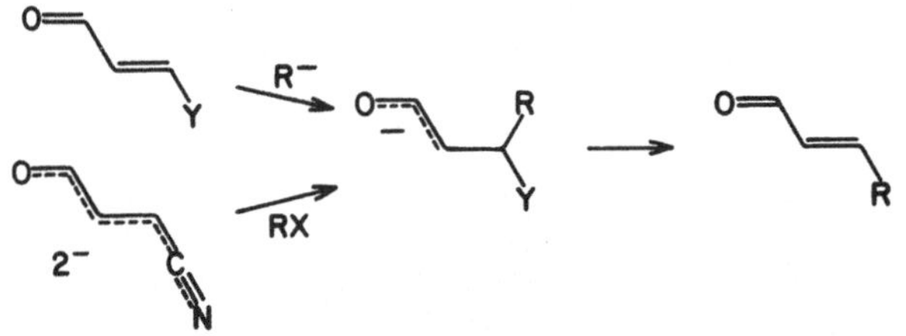

(6-9) [610] (Y = OAc),

(6-10) [612] (Y = CN)

RO$^-$ [183] and R$_2$N$^-$ [210] are known as leaving groups in reactions of type (6-2a). When it is desired to avoid elimination but these groups leave too rapidly, dianions *116* and *117* have been used successfully [613]. It should be noted that when *lithium* is used as the counterion, *116* will eliminate Li$_2$O at 25 °C; this has been profitably used in the conversions of α-haloketones (6-11) and epoxides (6-12) into alkenes.

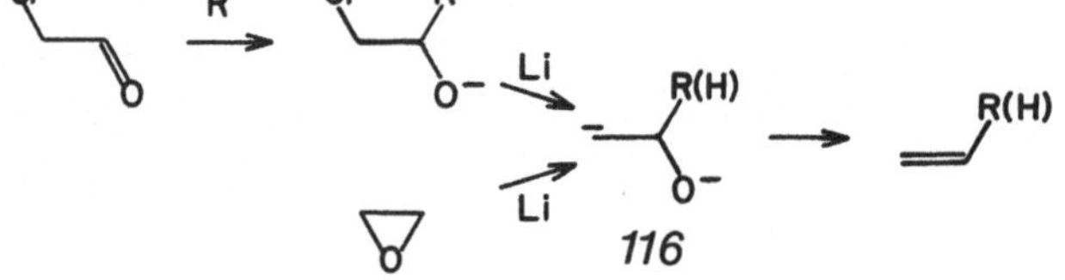

(6-11) [614]

(6-12) [615]

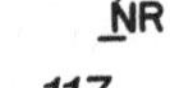

117

58

A type of reaction which probably involves β-elimination of RCO_2^- is (5-46), which presumably proceeds as shown in (6-13) (W = CO_2R). The very useful olefin-forming eliminations of R_3SiO^- [520], e.g. (6-14), probably also involve β-elimination, possibly in some cases as depicted.

$$ \text{(6-13) [544]} $$

$$ \text{(6-14) [616]} $$

The last step in a nucleophilic aromatic substitution by the addition-elimination mechanism (6-15) can be considered as a β-elimination of Y^-; the commonest leaving groups are halides. Groupings in the o- and p-positions of *118* which stabilize the intermediate cyclohexadienyl anion *119* are required for the reaction to proceed well.

$$ \text{(6-15) [617]} $$

118   *119*

Many carbanions have been observed to lose β-hydride, either by simply ejecting metal hydride (6-16) or by transferring hydride to an acceptor such as a trialkylborane [618, 619] or ketone (6-17). Cyclohexadienyllithium, which gives the highly stable substance benzene when hydride is lost, eliminates rapidly at 0 °C, but ethyllithium decomposes to lithium hydride and ethylene at 100 °C and ethylpotassium gives potassium hydride and ethylene with a half-life of 1 day at 25 °C [125]. The reduction of ketones (6-17) by Grignard reagents [620] and alkyllithiums [621], which can be an annoying side reaction during addition, especially with Grignard reagents, apparently involves hydride transfer from a simple alkyl carbanion. Reactions (6-18) and (6-19), which occur when the hydrocarbons are metallated, probably have as key steps the ejection of hydride from an intermediate di- or possibly trianion.

$$ \xrightarrow{\text{>0°C}} \quad + \quad H^- \qquad \text{(6-16) [275]} $$

$$ \text{(6-17) [620, 621]} $$

(6-18) [622]

(6-19) [623]

Ejection of a β-alkyl group from a carbanion is not usually as facile as hydride elimination, but in the case of cyclohexadienyl anions in which there is no good alternative (6-20) it is observed on heating; as expected, the more stable carbanion (methyl) is ejected in the example shown. *n*-Butyl-potassium gives some ethylene and ethylpotassium on standing at 25 °C [125].

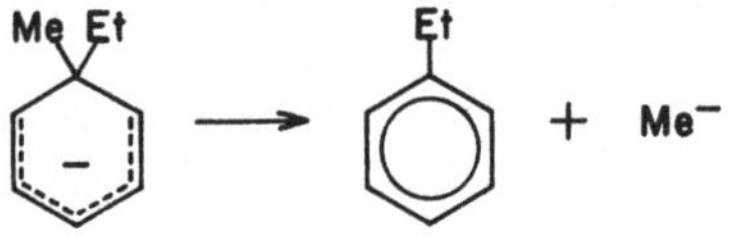

+ Me⁻ (6-20) [624]

The elimination of HX from vinyl halides with strong base has long been a good way to prepare alkynes; (6-2b) may be a key step in many of these reactions. Other leaving groups, e.g., ArSO₂⁻, can be used as well (6-21) [423]. An intriguing alkyne synthesis of unknown mechanism in which three hydrides are eliminated is (6-22) [625].

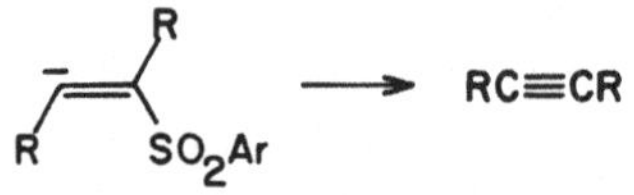

$\longrightarrow$ $RC{\equiv}CR$ (6-21)

+ 4 Li $\longrightarrow$ $LiC{\equiv}CR$ + 3 LiH (6-22)

3. γ-Eliminations

Reactions (6-23) to (6-25) show the use of (6-3) to make 3-membered rings; the latter two are termed Ramberg-Bäcklund reactions.

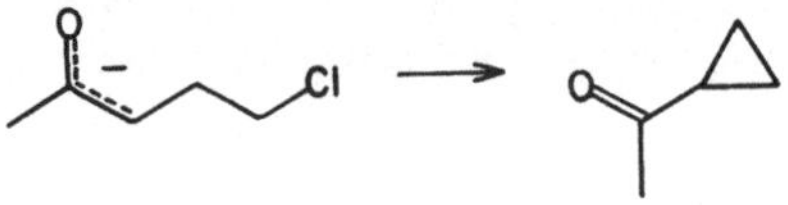

(6-23) [626]

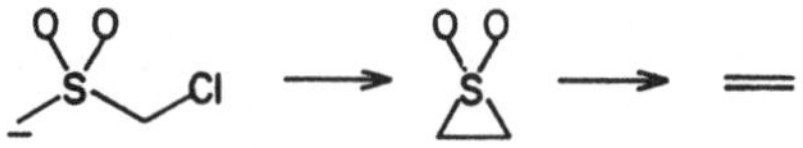

(6-24) [627]

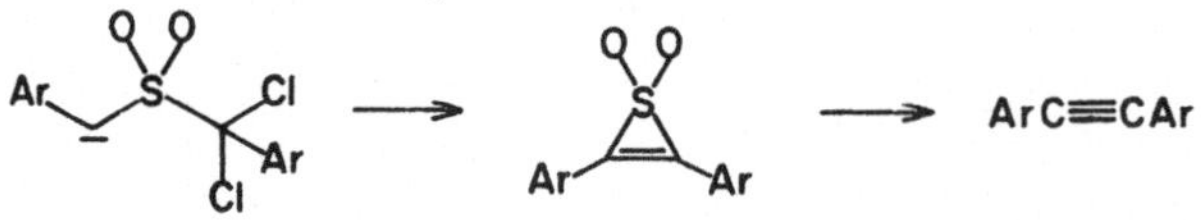

$\longrightarrow$ $ArC{\equiv}CAr$ (6-25) [628]

4. δ-Eliminations

The most useful type of reaction which is formally in this category is the Shapiro reaction (6-4) [629]. The vinyl anion can be reacted usefully with a variety of electrophiles [630–634], e.g., as in (6-26). This reaction has been used for carbonyl transposition as in (6-27) and for the synthesis of aryl-acetylenes (6-28).

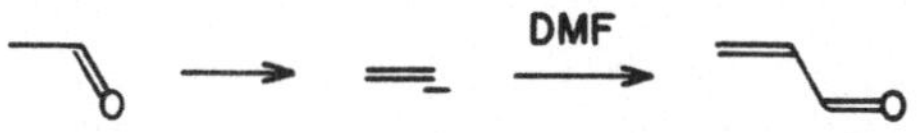

(6-26) [634]

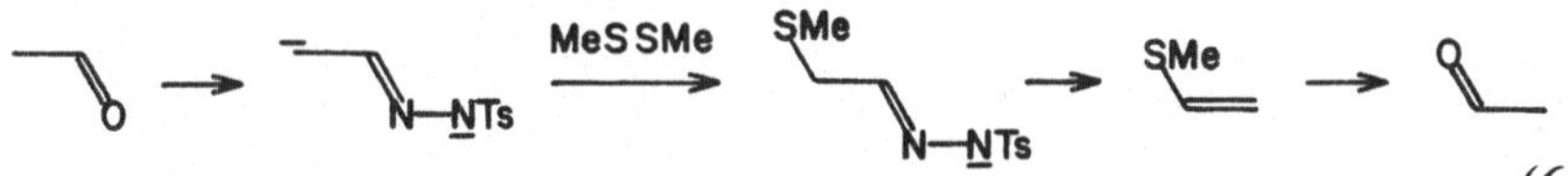

(6-27) [635]

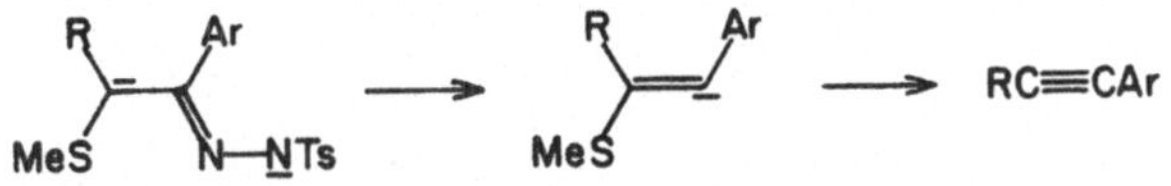

(6-28) [636]

5. Cycloeliminations

Cyclopentyl type anions *120–124* (unlike the corresponding cyclohexyl anions) undergo rapid ring-openings to allyl type anions *125–129* and ethylene or ethylene analogs (6-29) to (6-33).

(6-29) [293]

120   125

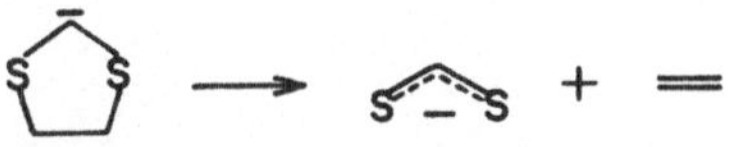

(6-30) [637]

121   126

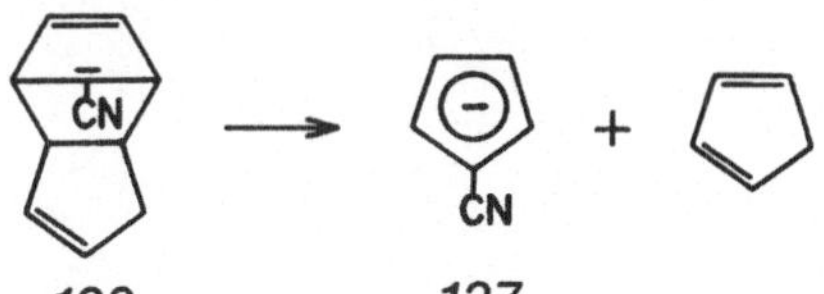

(6-31) [638]

122   127

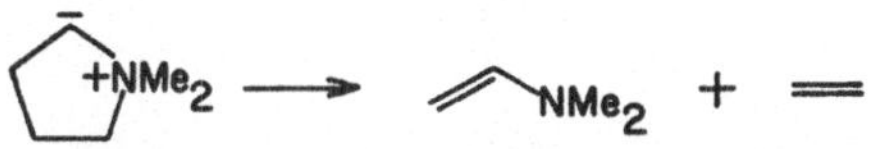

(6-32) [639]

123   128

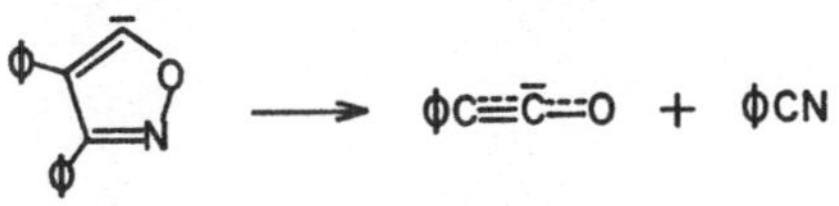

(6-33) [640]

124   129

π-cyclopentenyl anion *130* undergoes a different type of reverse [4+2] cycloaddition to give the same products observed from isomeric anion *122* (6-34).

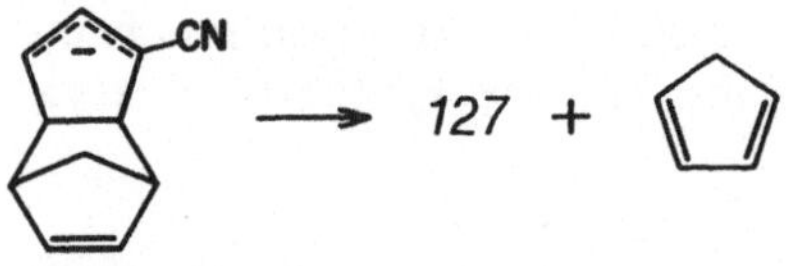

(6-34) [638]

130

VII. Oxidations

One-electron oxidation of monocarbanions leads to carbon radicals and two-electron oxidation gives carbonium ions [641]. In most of the oxidations below, the mechanism is not known (e.g., do oxidative coupling products arise from radical dimerization, combination of a cation with an anion, or nucleophilic displacement involving R^- and RY?), though progress is being made on some of the mechanisms [642, 643]. Again, the parallelism between base strength and ease of oxidation of carbanions should be noted [346].

1. Oxidations to Hydroperoxides, Alcohols, and Ketones

The reaction of O_2 with carbanions followed by quenching with protons (7-1) can give hydroperoxides in good yield in favorable cases [644, 645]. As some hydroperoxides are potentially explosive, they are usually reacted further without isolation; e.g., if the alcohol is desired, they can be reduced with a reagent such as sodium sulfite [646] or a trialkyl phosphite [647]. Alcohols can also be prepared via hydroperoxy molybdenum complexes (7-2) and alkylboranes (7-3). Oxygen can catalyze rotations about carbon-carbon bonds in allyl anions through reversible one-electron transfer [71].

$$ \begin{array}{cc} & (7\text{-}1) \\ & (7\text{-}2)\ [648] \\ & (7\text{-}3)\ [340, 341] \end{array} $$

Ketones can be prepared from special carbanion types *131–133* by elimination reactions which follow the oxygenation (7-4 to 7-6).

VII. Oxidations

$$\underset{131}{\overset{R}{\underset{R}{\diagdown}}\!\!\!\!-CN} \xrightarrow[\text{2) Sn}^{2+}]{\text{1) O}_2} \underset{R}{\overset{R}{\underset{OH}{\diagdown}}\!\!\!\!-CN} \qquad (7\text{-}4)\ [649]$$

$$\downarrow -HCN$$

$$\underset{132}{\overset{R}{\underset{R}{\diagdown}}\!\!\!\!-NO_2} \xrightarrow{{}^1O_2} \underset{R}{\overset{R}{\diagup}}\!\!\!=O \qquad (7\text{-}5)\ [650]$$

$$\downarrow -CO_2,\ -H_2O$$

$$\underset{133}{\overset{R}{\underset{R}{\diagdown}}\!\!\!\!-CO_2^-} \xrightarrow{O_2} \underset{R}{\overset{R}{\underset{OOH}{\diagdown}}\!\!\!\!-CO_2H} \qquad (7\text{-}6)\ [651]$$

2. Oxidative Couplings

The simple coupling reaction (7-7) works well only in certain cases. One of the most useful is the Eglinton reaction (7-8), which consists of the oxidation of acetylenic anions with ferric ion and goes very well indeed. Resonance-stabilized carbanions *134–136*, often containing carbanion-stabilizing groups W, have been oxidatively coupled with a variety of oxidizing agents to give products such as 1,2-dinitro compounds (7-9), 1,4-diketones (7-10), 1,4-diimines (7-11), and 1,5-dienes (7-12 and 7-13). Useful mixed couplings have been reported [655].

$$R^- \xrightarrow{[O]} RR \qquad (7\text{-}7)$$

$$RC\!\equiv\!C^- \xrightarrow{Fe^{3+}} RC\!\equiv\!CC\!\equiv\!CR \qquad (7\text{-}8)\ [652]$$

$$\underset{134}{\overset{}{\diagup}\!\!\!\!-W} \xrightarrow{[O]} \underset{W\quad W}{\diagup\!\!\!\diagdown} \qquad
\begin{array}{cc} \underline{W} & \underline{[O]} \\ -NO_2 & I_2 \\ -(C\!=\!O)R & Cu^{2+} \\ -(C\!=\!NR)R & Cu^+ \end{array}
\qquad \begin{array}{l}(7\text{-}9)\ [653]\\(7\text{-}10)\ [654, 655]\\(7\text{-}11)\ [656]\end{array}$$

$$\underset{135}{\overset{W\qquad W}{\underset{W\qquad W}{\diagdown\!\!\!-\!\!\!\diagup}}} \xrightarrow{Ce^{4+}} \qquad (7\text{-}12)\ [657]$$

$$\underset{136}{\diagup\!\!\!\diagdown\!\!\!\diagup} \xrightarrow{Zn^{2+}} \qquad + \qquad (7\text{-}13)\ [658]$$

3. Dianion Oxidations

Oxidation of dianions *137—144* leads to the products shown in (7-14) to (7-22); all are two-electron oxidation products except for *145–147*, which presumably arise by dimerization of the anion radical formed by one-electron oxidation of dianions *140–142*. The first two oxidations (7-14 and 7-15) are very important in biological systems. All of these 2-electron oxidations can represent the second step in a general dehydrogenation process in which the first step consists of abstracting two protons, and the second, two electrons [661–666].

(7-14)

137

(7-15)

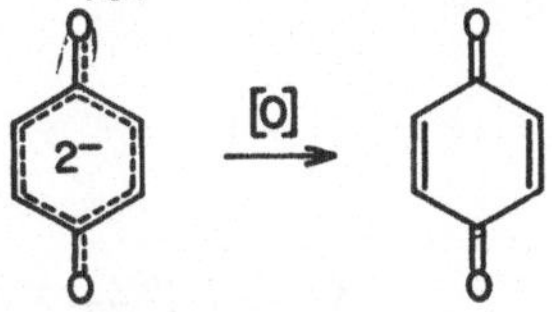

138

(7-16) [73]

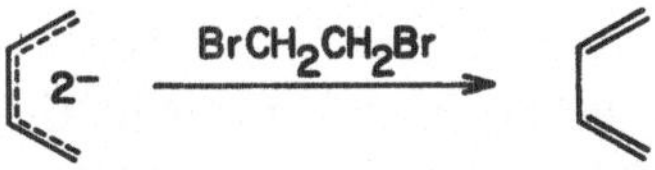

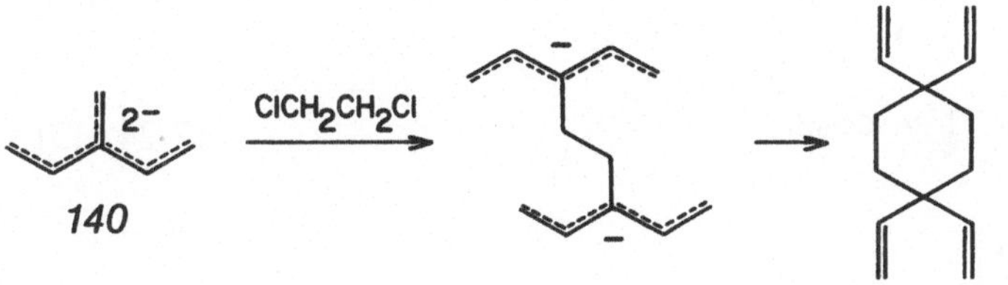

139

(7-17) [659]

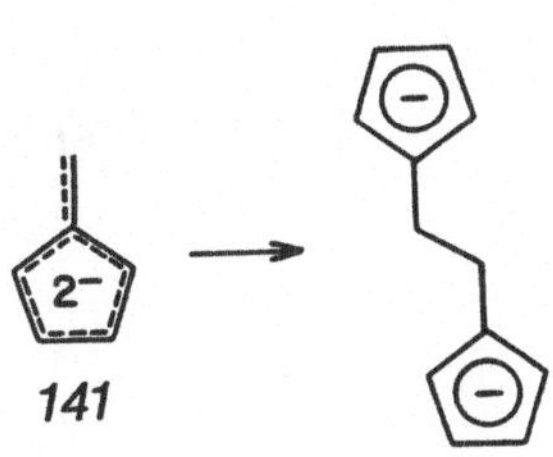

140

145

(7-18) [660]

141

146

65

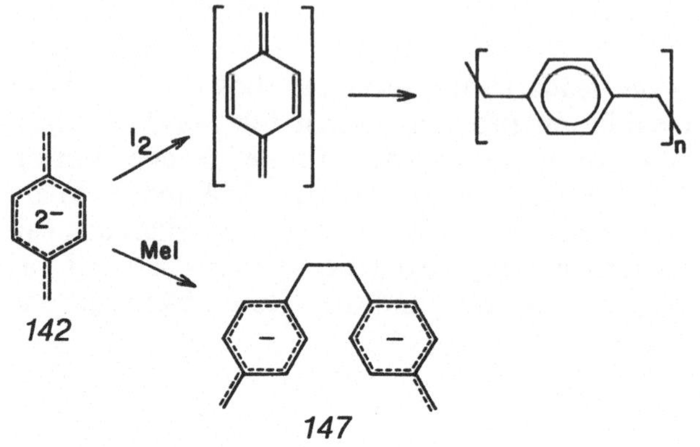

(7-19) [661]

(7-20) [661]

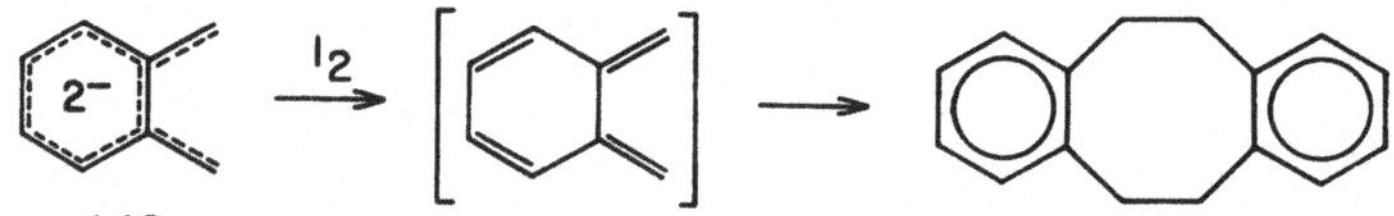

(7-21) [661]

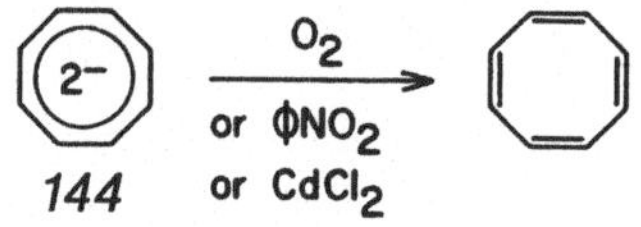

(7-22) [622, 662]

When the product of two-electron oxidation would be a diradical, the isolated products can usually be thought of as arising from intra- or intermolecular coupling of this diradical (7-23 to 7-27).

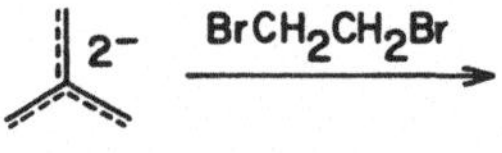

(7-23) [73]

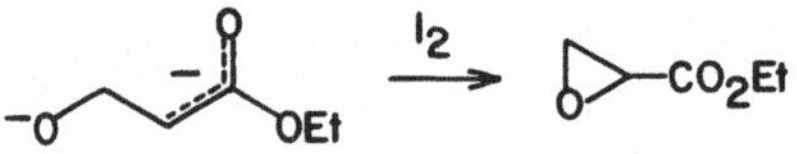

(7-24) [518]

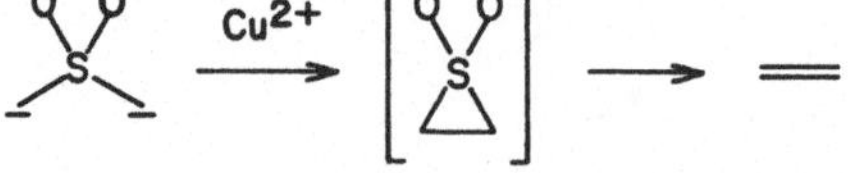

(7-25) [667]

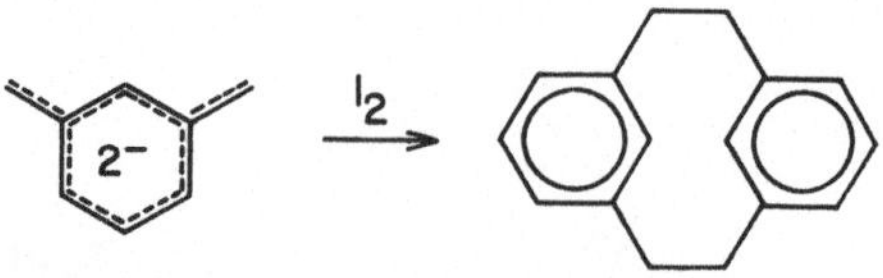

(7-26) [661]

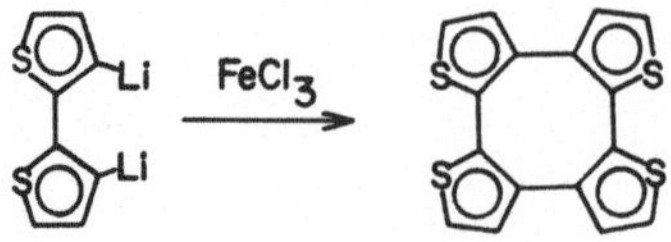

(7-27) [668]

VIII. Rearrangements

The various types of simple intramolecular rearrangements described in Sections 2 to 5 below are often combined in complex sequences, some examples of which are given in Section 6.

1. Intermolecular Rearrangements

The equilibrium (8-1) in which the counterion M^+ becomes more and less tightly associated with the carbanion R^- can account for stereochemical changes in the carbon species present: (a) σ carbanions, in which the tight structure R—M is favored, can have the configurations at their carbons inverted *via* the loose form R^-; this is most easily detected by racemization in a case where the charge-bearing sp^3 hybridized carbon is the only chiral center (Chapter II, section 1, part b); (b) π carbanions, in which the loose form is favored, can undergo rotations about partial double bonds which essentially become single bonds in the tight form (8-2); if R = vinyl, this could represent a change from W- to U-shaped pentadienyl anion. Rotations can of course occur about partial double bonds in the delocalized carbanion itself, independent of M^+, but this seems less likely than the above. π anions have been geometrically isomerized photochemically as well as thermally [669].

$$R^- \ + \ M^+ \ \rightleftarrows \ RM \tag{8-1}$$

$$\tag{8-2}$$

More drastic rearrangement of a π carbanion can occur through the equilibrium (8-1) if M = H; the proton can come off a different allylic carbon, causing a shift of the allyl anion system down the chain in addition to any stereochemical changes (8-3). This sequence is a vital part of base-catalyzed alkene isomerizations, which are generally run under conditions where the alkenes are much more stable than the allyl anions, and the *alkenes* are thus equilibrated rather than the allyl anions [487, 670, 671]. Sometimes, however, the *anions* themselves are equilibrated, as when the kinetic dimetallation product *148* is converted to the thermodynamically more stable isomer *1·* on standing at 25 °C (8-4) [672]; for other examples, see [282, 283].

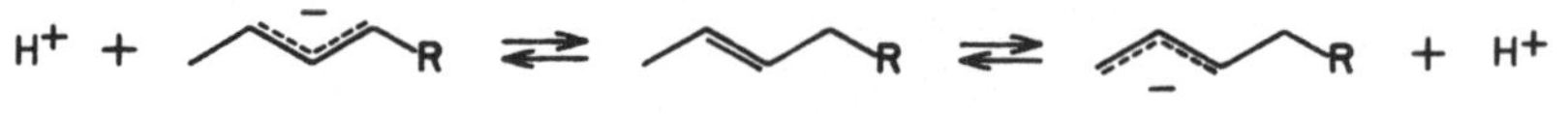

$$H^+ + \ \diagdown\!\!\diagup\!\!\diagdown\!\!R \ \rightleftarrows \ \diagdown\!\!\diagup\!\!\diagdown\!\!R \ \rightleftarrows \ \diagdown\!\!\diagup\!\!\diagdown\!\!R \ + \ H^+ \qquad (8\text{-}$$

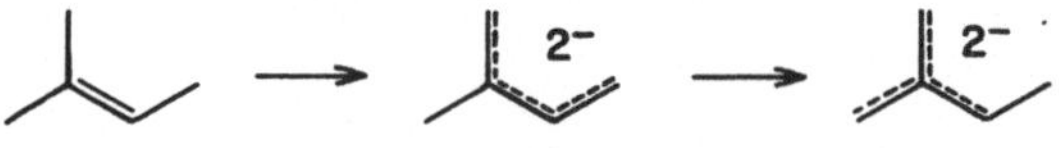

148   149

Acetylenes and allenes can similarly be equilibrated efficiently by ba (8-5). Again, the bases used are usually not sufficiently strong to provi(high concentrations of the anions, *unless* R or R′ = H, in which case tl final step consists of abstraction of the terminal acetylenic proton to gi' the acetylenic anion, e.g., (8-6).

$$RCH_2C\equiv CR' \ \rightleftarrows \ RCH\!=\!\bar{C}\!\equiv\!CR' \ \rightleftarrows \ RCH\!=\!C\!=\!CHR' \ \rightleftarrows \ R\bar{C}\!\equiv\!C\!=\!CHR' \ \rightleftarrows \ RC\equiv CCH_2R'$$

(8-5) [67

$$Me(CH_2)_nC\equiv CR \ \longrightarrow \ {}^-C\equiv C(CH_2)_{n+1}R$$

(8-6) [673, 67

2. Intramolecular Additions

5-Membered rings are kinetically favored over 6; e.g., reaction (8-7) is fast than (8-8) by a factor of 2800 [675]. The preference for an anion of the ty] *150* to form a 5- rather than a 6-membered ring is borne out in other systen [73, 676, 677].

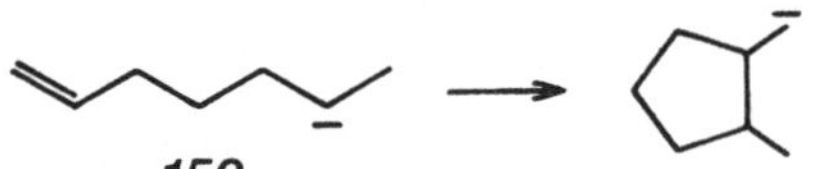

150

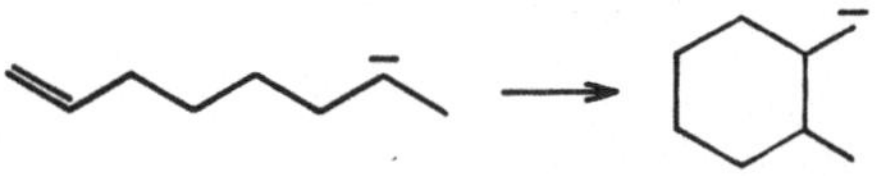

The most important intramolecular additions are probably ring-formir aldol condensations, e.g. (8-9). These are usually equilibrium-controll(reactions in which only 5- and 6-membered rings are observed.

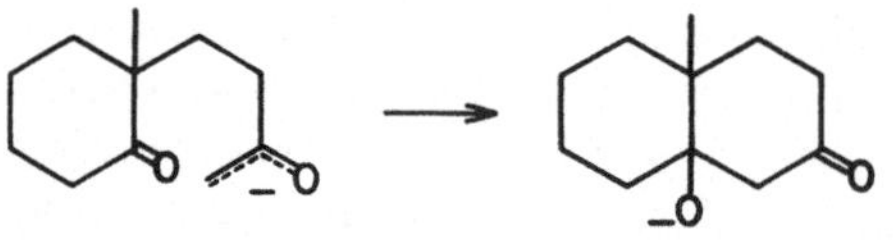

(8-9) [67

Reaction (8-10) shows the formation of cyclopropane and cyclobutane rings by intramolecular additions; in this case the energy to make such thermodynamically unfavored ring sizes is provided by the opening of an epoxide and the shift of charge from carbon to oxygen. Li^+ at room temperature favored the former product and Mg^{2+} at -70 °C the latter.

$$(8\text{-}10) \; [679]$$

3. Intramolecular Eliminations

These reactions are simply the reverse of the reactions in Section 2 [680]. The driving force is often relief of ring strain (8-11), or leaving group stabilization as in (8-12) and reverse aldols (8-9), or both.

$$(8\text{-}11) \; [681]$$

$$(8\text{-}12) \; [682]$$

Vinyl (8-13) and heteroaromatic (8-14 and 8-15) anions can also undergo ring opening, even though an aromatic ring is destroyed in the process in the latter cases. Benzylic-type anions (8-16 and 8-17) also undergo cleavage of the aromatic ring [688].

$$(8\text{-}13) \; [683]$$

$$(8\text{-}14) \; [684]$$

$$(8\text{-}15) \; [685]$$

69

(8-16) [686]

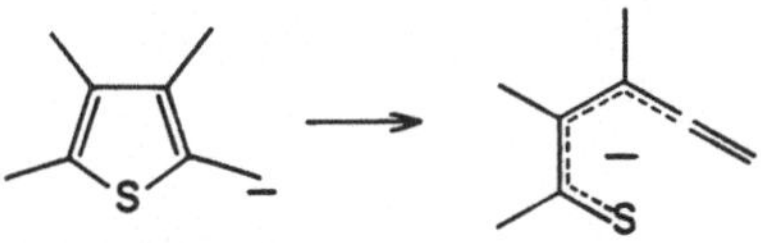

(8-17) [687]

4. Sigmatropic Carbanion Rearrangements

In sharp contrast to carbonium ions, carbanions undergo [1,2]sigmatropic shifts very rarely if ever, in accordance with the Woodward-Hoffmann rules [689]. 1,2-Aryl shifts occur, but by addition-elimination rather than concertedly (see Section 6). Stevens (8-18, 8-19) and Wittig (8-20) rearrangements involve 1,2-alkyl shifts but probably proceed by diradical mechanisms [692, 695].

$$Me_3\overset{+}{N}CH_2^- \longrightarrow Me_2NEt \qquad\qquad (8\text{-}18)\ [690\text{-}692]$$

$$Me_2\overset{+}{S}\overline{C}HR \longrightarrow MeSCHRMe \qquad\qquad (8\text{-}19)$$

$$MeO\overline{C}HR \longrightarrow {}^-OCHRMe \qquad\qquad (8\text{-}20)\ [693\text{-}695]$$

Suprafacial [1,4]proton shifts in allyl anions (8-21) are permitted thermally but have not been observed due to the thermal instability of allyl anions.

(8-21)

On the other hand, some thermal antarafacial [1,6]sigmatropic proton shifts in pentadienyl anions (8-22) occur just above room temperature [696, 697] *via* a coiled transition state (*151*). Steric hindrance (R's = alkyl in *151*) slows the rearrangement; each R = Me up to two (the maximum tried) increases the activation energy by several Kcal/mole. The photochemical suprafacial counterpart of this reaction was found in (8-23) [696].

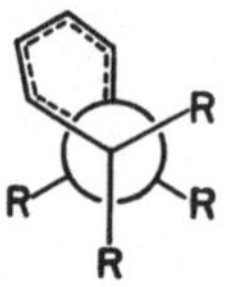

151

(8-22)

(8-23)

Efforts to observe thermal [1,8]sigmatropic proton shifts in heptatrienyl anions failed.

Many [2,3]sigmatropic rearrangements of heteroatom-containing carbanions are known [698–703]. The usual driving force for these reactions is the transfer of charge from carbon to a heteroatom, which can be N (8-24, 8-25), O (8-26, 8-27), or S (8-28). (8-26) is a β,γ-unsaturated ketone preparation, and (8-27) provides a means of getting only *ortho* substitution in the acylation of an alkylbenzene.

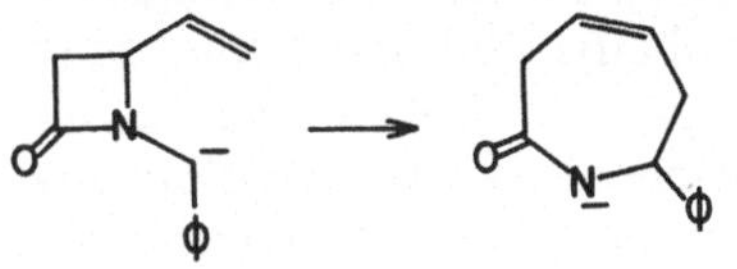

(8-24) [698]

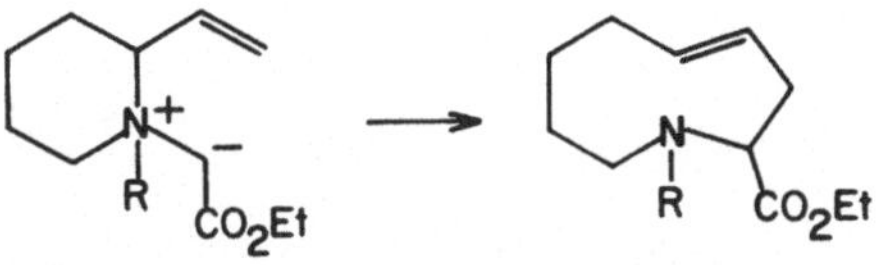

(8-25) [699]

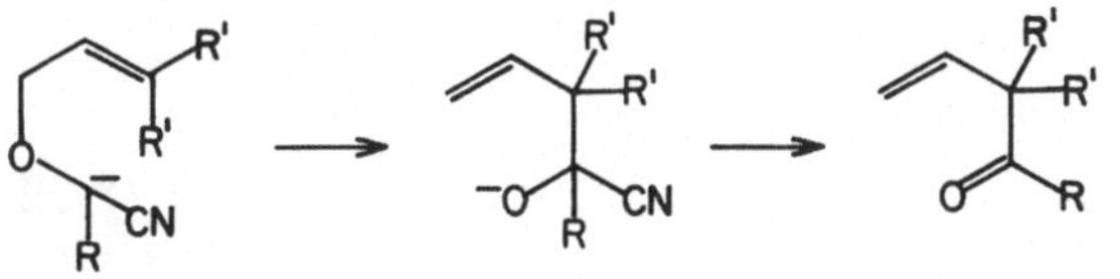

(8-26) [700]

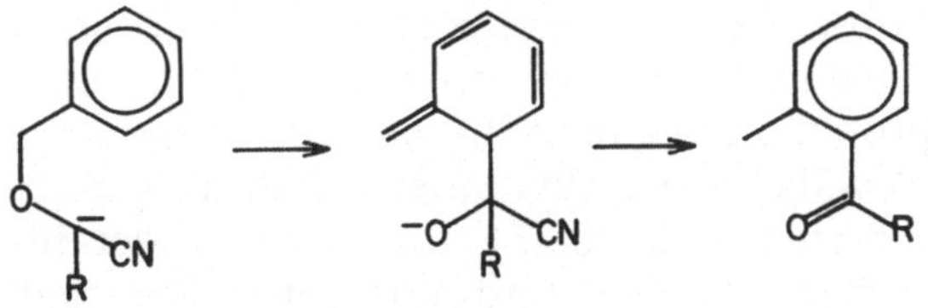

(8-27) [701]

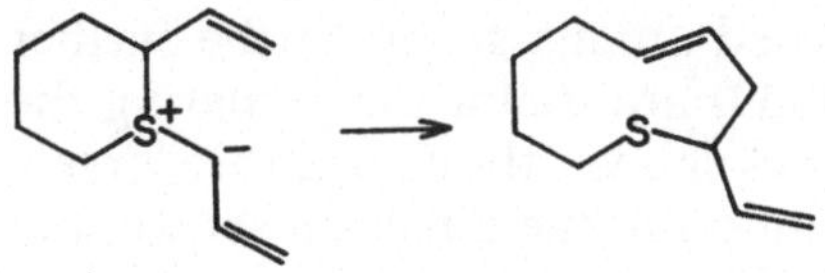

(8-28) [703]

A [3,3]sigmatropic rearrangement (8-29) in which a thioenolate anion rearranges has been observed [704]; the driving force is presumably the transferring of partial negative charge from carbon to the much more electronegative sulfur.

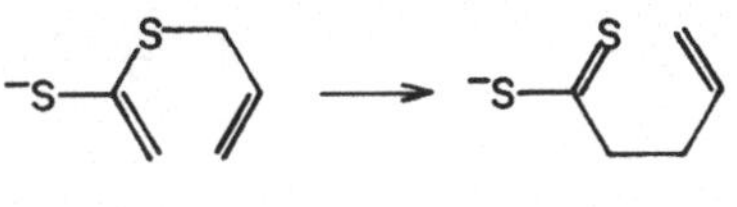

(8-29)

5. Electrocyclic Carbanion Rearrangements

In the all-carbon unsubstituted systems, the thermal equilibria in the cyclopropyl ⇆ allyl (8-30), cyclopentenyl ⇆ pentadienyl (8-31), and cycloheptadienyl ⇆ heptatrienyl (8-32) anion rearrangements probably lie in the directions shown; this has only been actually demonstrated in the last case, where the energy barrier is especially low and the reaction proceeds rapidly just above −30 °C [314, 705, 706]. The last two examples require the open system to be coiled rather than extended as shown, but the rotation barriers about the inner carbon-carbon bonds in pentadienyl [79] and heptatrienyl [314] anions are known to be low enough that this is no problem. In the first two cases, the open anions have not been observed to close; cyclopropyl anion can be prepared and reacted without opening [55, 56], whereas cyclopentenyl anion has resisted preparation.

$$\text{(8-30)}$$

$$\text{(8-31)}$$

$$\text{(8-32)}$$

These reactions have all been observed in systems containing substituents (8-33 to 8-35) and heteroatoms (8-36 to 8-39) which can change the position of equilibrium [707]. The cyclopropyl anions in (8-33) and (8-34) were not observed at 25 °C but opened rapidly to the allyl anions. Putting a pentadienyl system in an 8-membered ring as in (8-35) strains it considerably (none of the shapes with the π system planar is unstrained) with the result that it closes to the cyclopentenyl system at 35 °C with a half life of 90 minutes. The positions of the electrocyclic equilibria in (8-36) and (8-37) are not known since under the reaction conditions the closed anions go on to the further products depicted. Reactions (8-38) and (8-39) are useful for preparing the dienolate and trienolate anions with the shapes shown; the charges are largely on oxygen in the products, driving the equilibria in the direction shown and causing high rotation barriers about the C2–C3 bonds which prevent equilibration with the fully extended shapes.

$$\text{(8-33) [708}$$

$$\text{(8-34) [709}$$

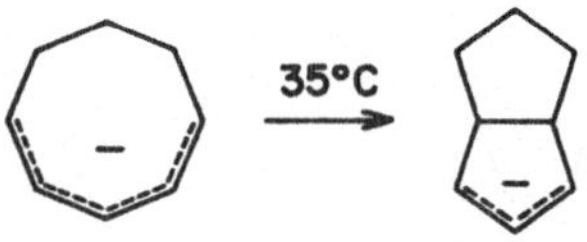

(8-35) [710]

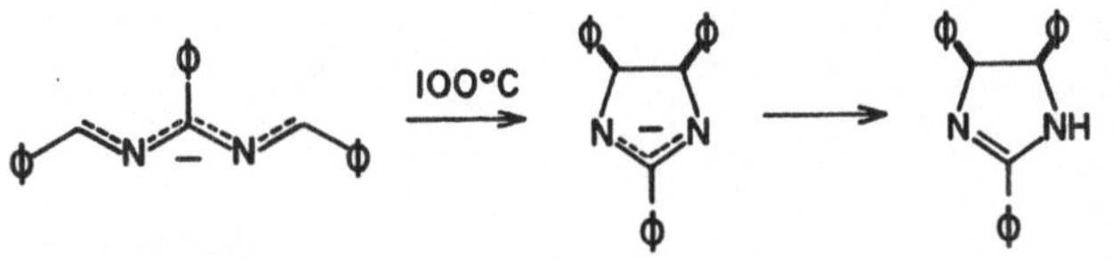

(8-36) [711]

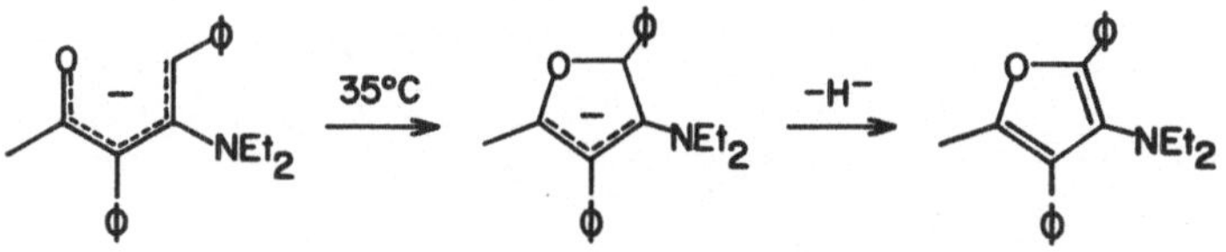

(8-37) [712]

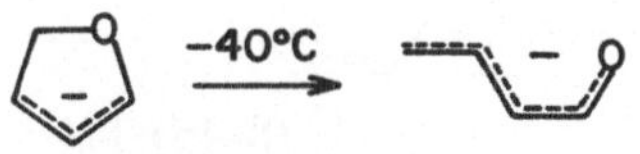

(8-38) [713, 293]

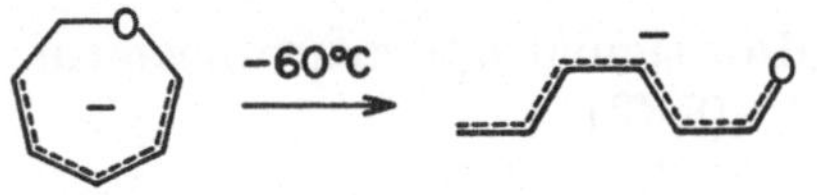

(8-39) [714]

A dianion electrocyclization is apparently involved in (8-40).

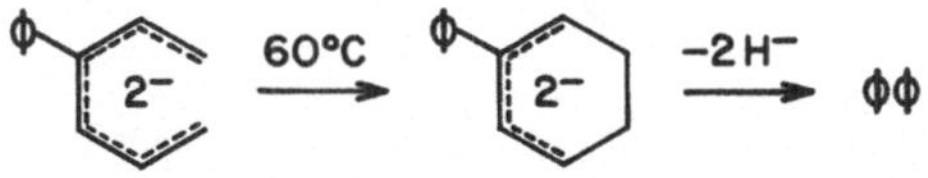

(8-40) [715]

6. Complex Intramolecular Rearrangements

Some sequences combining additions and eliminations are shown in (8-41) to (8-45).

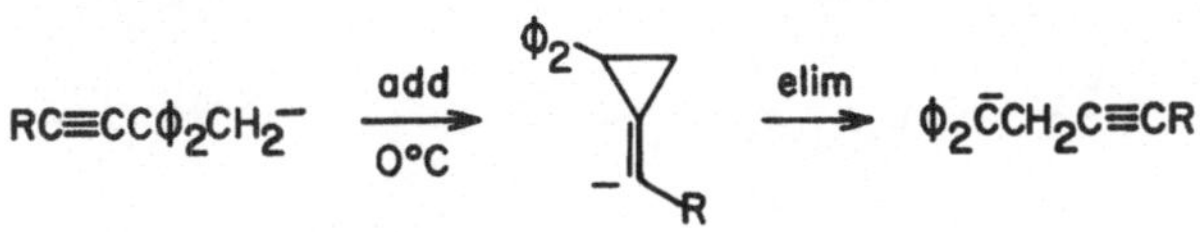

(8-41) [716]

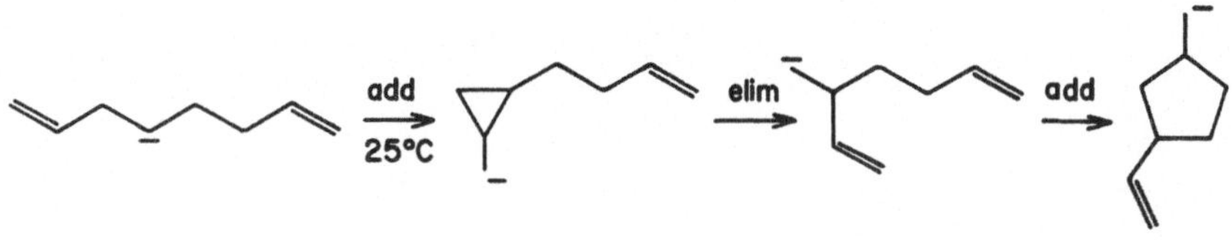

(8-42) [717]

73

$$\text{(8-43) [715}$$

$$\text{(8-44) [718}$$

$$\text{(8-45) [719}$$

Combinations including electrocyclic and/or sigmatropic rearrangement as well as other types are shown in (8-46) to (8-48).

$$\text{(8-46) [720}$$

$$\text{(8-47) [696}$$

$$\text{(8-48) [721}$$

IX. Carbanion Equivalents

When a carbanion is inherently unstable or reacts in the wrong way, it may well prove possible to use a several step route involving a "carbanion equivalent" to accomplish the desired synthetic goal. Acyl anions, for example, lose CO at 25 °C, but hundreds of acyl anion equivalents have been devised which serve the same purpose. The carbanion equivalents with reference numbers below 722 are mentioned in the text and can be located in the text from the reference number.

C_1	$NH_2CH_2^-$	178, 571, 722
	$RNHCH_2^-$	585–589
	$\Phi_3\overset{+}{P}CH_2^-$	191, 216, 443–449, 723, 724
	$ROCH_2^-$	192
	XCH_2^-	725, 726
	$H(C{=}O)^-$	193, 194, 206, 727–732
	$HO(C{=}O)^-$	207, 728, 733
C_2	RCH_2^-	351, 734
	$HOCH_2CH_2^-$	613
	$H(C{=}O)CH_2^-$	141, 454, 542, 727, 735
	$NCCH_2^-$	543
	$HO(C{=}O)CH_2^-$	552–557, 565, 566
	$HO(C{=}O)\overline{C}HNH_2$	736
	$HO\overline{C}HR$	182
	$R(C{=}O)^-$	183, 199, 202, 210, 343, 352, 455, 571, 592, 727, 728, 737–749
C_3	$HOCH_2CH_2CH_2^-$	750
	$H(C{=}O)CH_2CH_2^-$	751–753
	$HO(C{=}O)CH_2CH_2^-$	754, 755
	$RNHCH_2CH_2^-$	613
	$CH_2{=}CHCH_2^-$	756, 757
	$R(C{=}O)CH_2^-$	418, 542, 576, 758
	$CH_2{=}CH\overline{C}HOH$	759, 760
	$R\overline{C}H{=}CH^-$	761, 762
	$HOCH_2CH{=}CH^-$	763
	$H(C{=}O)CH{=}CH^-$	764, 765
	$CH_2{=}\overline{C}COOH$	354, 766–770

IX. Carbanion Equivalents

C_4 $R(C=O)CH_2CH_2^-$ 771
    $R(C=O)CH=CH^-$ 771, 772
    $CH_2=CH\bar{C}=CH_2$ 773
    $R(C=O)\bar{C}=CH_2$ 774
    $H(C=O)CH_2CH_2(C=O)^-$ 775

C_5 $CH_2=CH(C=O)CH_2CH_2^-$ 550
    $HOCH_2RC=CHCH_2^-$ 757
    $^-CH_2(C=O)CH=CHCH_2^-$ 776
    $CH_2=C(CH_2^-)CH=CH_2$ 773
    $H(C=O)CH_2CH_2(C=O)CH_2^-$ 777
    $H(C=O)CH_2CH_2CH=CH^-$ 778

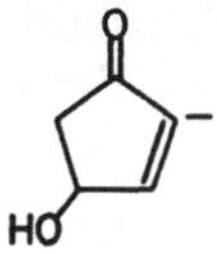

                                    779

    $H(C=O)CH_2CH_2CH_2(C=O)^-$ 775
    $R(C=O)CH_2CH_2(C=O)^-$ 780

C_6 $HO(C=O)CH_2CH_2CH_2CH_2CH_2^-$ 781, 782
    $HO(C=O)CH_2CH_2CH_2(C=O)CH_2^-$ 781, 782

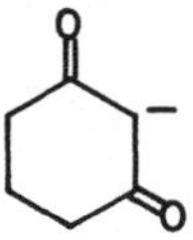

                                    783

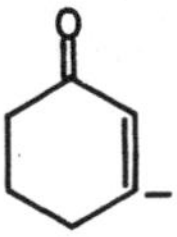

                                    784

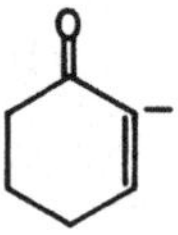

                                  540, 568, 569, 785

X. Summary

Carbanions owe their importance as intermediates in synthesis to the great variety of bond-forming reactions which they undergo which lead to bonds from carbon to every element except the noble gases. Unlike many carbonium ions, most carbanions do not undergo rearrangements but in many cases react in substitution and addition reactions in high yield. Changing the cation and solvent can greatly affect the reaction, even without employing transition metals which add a further dimension to their reactivity.

Much further work on the structures of carbanion salts and on the mechanisms of their reactions is necessary before their utility in synthesis can be fully exploited.

XI. References

1. Cram, D. J.: Fundamentals of Carbanion Chemistry, New York: Academic Press 1965
2. Tsuji, J.: Organic Synthesis by Means of Transition Metal Complexes, Berlin, Heidelberg, New York: Springer 1975
3. Tsuji, J.: Organic Synthesis with Palladium Compounds, Berlin, Heidelberg, New York: Springer 1980
4. Ellison, G. B., Engelking, P. C., Lineberger, W. C.: J. Am. Chem. Soc. *100*, 2557 (1978)
5. Whitesides, G. M., Witanowski, M., Roberts, J. D.: J. Am. Chem. Soc. *87*, 2854 (1965)
6. McKeever, L. D., Waack, R.: Chem. Commun. *1969*, 750
7. Guggenberger, L. J., Rundle, R. E.: J. Am. Chem. Soc. *90*, 5375 (1968)
8. Zerger, R. P., Stucky, G. D.: Chem. Commun. *1973*, 44
9. Weiss, E., Hencken, G.: J. Organometal. Chem. *21*, 265 (1970)
10. Dietrich, H.: J. Organometal. Chem. *205*, 291 (1981)
11. Zerger, R., Rhine, W., Stucky, G.: J. Am. Chem. Soc. *96*, 6048 (1974)
12. Glaze, W. H., Freeman, C. H.: J. Am. Chem. Soc. *91*, 7198 (1969)
13. Brown, T. L.: Accts. Chem. Res. *1*, 23 (1968)
14. Quirk, R. P., Kester, D. E.: J. Organometal. Chem. *127*, 111 (1977)
14a. Amstutz, R., Dunitz, J. D., Seebach, D.: Angew. Chem., Int. Ed. Engl. *20*, 465 (1981)
15. Graham, G., Richtsmeier, S., Dixon, D. A.: J. Am. Chem. Soc. *102*, 5759 (1980)
16. Thoennes, D., Weiss, E.: Chem. Ber. *111*, 3157 (1978)
17. Köster, H., Thoennes, D., Weiss, E.: J. Organometal. Chem. *160*, 1 (1978)
18. Dietrich, H., Rewicki, D.: J. Organometal. Chem. *205*, 281 (1981)
19. Arora, S. K. et al.: J. Am. Chem. Soc. *97*, 6271 (1975)
20. Patterman, S. P., Karle, I. L., Stucky, G. D.: J. Am. Chem. Soc. *92*, 1150 (1970)
21. Rhine, W. E., Stucky, G. D.: J. Am. Chem. Soc. *97*, 737 (1975)
22. Brooks, J. J., Rhine, W., Stucky, G. D.: J. Am. Chem. Soc. *94*, 7346 (1972)
23. Rhine, W. E., Davis, J. H., Stucky, G.: J. Organometal. Chem. *134*, 139 (1977)
24. Brooks, J. J., Rhine, W., Stucky, G. D.: J. Am. Chem. Soc. *94*, 7339 (1972)
25. Rhine, W. E., Davis, J., Stucky, G.: J. Am. Chem. Soc. *97*, 2079 (1975)
26. Brooks, J. J., Stucky, G. D.: J. Am. Chem. Soc. *94*, 7333 (1972)
27. Bladauski, D. et al.: Angew. Chem., Int. Ed. Engl. *16*, 474 (1977)
27a. Amstutz, R. et al.: Angew. Chem., Int. Ed. Engl. *19*, 53 (1980)
28. Weiss, E., Plass, H.: Chem. Ber. *101*, 2947 (1968)
29. Aoyagi, T. et al.: J. Organometal. Chem. *175*, 21 (1979)
30. Köster, H., Weiss, E.: J. Organometal. Chem. *168*, 273 (1979)

31. Weiss, E., Sauermann, G.: Chem. Ber. *103*, 265 (1970)
32. Goldberg, S. Z. et al.: J. Am. Chem. Soc. *96*, 1348 (1974)
33. Zerger, R., Rhine, W., Stucky, G. D.: J. Am. Chem. Soc. *96*, 5441 (1974)
34. Weiss, E., Köster, H.: Chem. Ber. *110*, 717 (1977)
35. Snow, A. I., Rundle, R. E.: Acta Cryst. *4*, 348 (1951)
36. Greiser, T. et al.: J. Organometal. Chem. *191*, 1 (1980)
37. Toney, J., Stucky, G. D.: Chem. Commun. *1967*, 1168
38. Spek, A. L. et al.: J. Organometal. Chem. *77*, 147 (1974)
39. Stucky, G., Rundle, R. E.: J. Am. Chem. Soc. *86*, 4825 (1964)
40. Atwood, J. L., Smith, K. D.: J. Am. Chem. Soc. *96*, 994 (1974)
41. Weiss, E.: J. Organometal. Chem. *2*, 314 (1964)
42. Weiss, E.: J. Organometal. Chem. *4*, 101 (1965)
43. Johnson, C., Toney, J., Stucky, G. D.: J. Organometal. Chem. *40*, C11 (1972)
44. Lewis, P. H., Rundle, R. E.: J. Chem. Phys. *21*, 986 (1953)
45. Schonberg, P. R. et al.: Organometallics *1*, 799 (1982)
46. Pauling, L.: The Nature of the Chemical Bond, 3rd Ed., Ithaca, New York: Cornell University Press 1960
47. West, P., Waack, R.: J. Am. Chem. Soc. *89*, 4395 (1967)
48. Seitz, L. M., Brown, T. L.: J. Am. Chem. Soc. *88*, 2174 (1966)
49. Berkowitz, J., Bafus, D. A., Brown, T. L.: J. Phys. Chem. *65*, 1380 (1961)
50. Wakefield, B. J.: Chemistry of Organolithium Compounds, Oxford: Pergamon Press 1974
51. Schröder, F. A.: Chem. Ber. *102*, 2035 (1969)
52. Normant, H., Maitte, P.: Bull. Soc. Chim. France *1956*, 1439
53. Curtin, D. Y., Koehl, W. J., Jr.: J. Am. Chem. Soc. *84*, 1967 (1962)
54. Bates, R. B., Cutler, R. S., Freeman, R. M.: J. Org. Chem. *42*, 4162 (1977)
55. Walborsky, H. M., Young, A. E.: J. Am. Chem. Soc. *86*, 3288 (1964)
56. Applequist, D. E., Peterson, A. H.: J. Am. Chem. Soc. *83*, 862 (1961)
57. Pellerite, M. J., Brauman, J. I. in Buncel, E., Durst, T.: Comprehensive Carbanion Chemistry, p. 55, New York: Elsevier 1980
58. Ziolo, R. F., Günther, W. H. H., Troup, J. M.: J. Am. Chem. Soc. *103*, 4629 (1981)
59. Buncel, E., Menon, B. in Buncel, E., Durst, T.: Comprehensive Carbanion Chemistry, p. 97, New York: Elsevier 1980
60. Fox, M. A.: Chem. Rev. *79*, 253 (1979)
61. Corset, J. in Buncel, E., Durst, T.: Comprehensive Carbanion Chemistry, p. 125, New York: Elsevier 1980
62. O'Brien, D. H. in Buncel, E., Durst, T.: Comprehensive Carbanion Chemistry, p. 271, New York: Elsevier 1980
63. Brownstein, S., Bywater, S., Worsfold, D. J.: J. Organometal. Chem. *199*, 1 (1980)
64. Fraenkel, G. et al.: J. Organometal. Chem. *197*, 249 (1980)
65. Yasuda, H., Yamauchi, M., Nakamura, A.: J. Organometal. Chem. *202*, C1 (1980)
66. Fraenkel, G., Hallden-Abberton, M. P.: J. Am. Chem. Soc. *103*, 5657 (1981)
67. Richey, H. G., Jr. in Zabicky, J.: The Chemistry of Alkenes *2*, p. 67, New York: Interscience, 1970
68. Thompson, T. B., Ford, W. T.: J. Am. Chem. Soc. *101*, 5459 (1979)
69. Neugebauer, W., Schleyer, P. v. R.: J. Organometal. Chem. *198*, C1 (1980)
70. O'Brien, D. H., Russell, C. R., Hart, A. J.: Tetrahedron Lett. *1976*, 37
71. Stähle, M., Hartmann, J., Schlosser, M.: Helv. Chim. Acta *60*, 1730 (1977)
72. Bates, R. B. et al.: J. Am. Chem. Soc. *96*, 5640 (1974)

XI. References

73. Bahl, J. J. et al.: J. Org. Chem. *41*, 1620 (1976)
74. Fujita, K. et al.: J. Organomet. Chem. *113*, 201 (1976)
75. Hoffmann, R., Olofson, R. A.: J. Am. Chem. Soc. *88*, 943 (1966)
76. Bingham, R. C.: J. Am. Chem. Soc. *98*, 535 (1976)
77. Bartmess, J. E. et al.: J. Am. Chem. Soc. *99*, 1976 (1977)
78. Hogen-Esch, T. E., Jenkins, W. L.: J. Am. Chem. Soc. *103*, 3666 (1981)
79. Bates, R. B., Gosselink, D. W., Kaczynski, J. A.: Tetrahedron Lett. *1967* 205
80. Staley, S. W., Dustman, C. K.: J. Am. Chem. Soc. *103*, 4297 (1981)
81. Klein, J., Brenner, S.: Tetrahedron *26*, 5807 (1970)
82. West, R.: Adv. Chem. Ser. *130*, 211 (1974)
83. Priester, W., West, R., Chwang, T. L.: J. Am. Chem. Soc. *98*, 8413 (1976)
84. Shimp, L. A. et al.: J. Am. Chem. Soc. *103*, 5951 (1981)
85. Klein, J., Becker, J. Y.: Chem. Commun. *1973*, 576
86. Jackman, L. M., Lange, B. C.: Tetrahedron *33*, 2737 (1977)
87. House, H. O., Prabhu, A. V., Phillips, W. V.: J. Org. Chem. *41*, 1209 (1976)
88. Fellmann, P., Dubois, J. E.: Tetrahedron Lett. *1977*, 247
89. Lynch, T. J. et al.: J. Org. Chem. *45*, 5005 (1980)
90. Lee, J. Y. et al.: J. Am. Chem. Soc. *103*, 6215 (1981)
90a. Amstutz, R. et al.: Helv. Chim. Acta *64*, 2622 (1981)
90b. Amstutz, R. et al.: Helv. Chim. Acta *64*, 2617 (1981)
91. Macintyre, W. M., Werkema, M. S.: J. Chem. Phys. *40*, 3563 (1964)
92. Allmann, R. et al.: Chem. Ber. *109*, 2208 (1976)
93. Baenziger, N. C., Hegenbarth, J. J., Williams, D. G.: J. Am. Chem. Soc. *85*, 1539 (1963)
94. Schröder, F. A., Weber, H. P.: Acta Crystallogr. *B31*, 1745 (1975)
95. Bright, D., Milburn, G. H. W., Truter, M. R.: J. Chem. Soc. (A) *1971*, 1582
96. Fenton, D. E., Nave, C., Truter, M. R.: J. Chem. Soc. (Dalton) *1973*, 2188
97. Bogdanovic, B., Krüger, C., Wermeckes, B.: Angew. Chem., Int. Ed. Engl. *19*, 817 (1980)
98. Jackman, L. M., Lange, B. C.: J. Am. Chem. Soc. *103*, 4494 (1981)
99. Lochmann, L., De, R., Trekoval, T.: J. Organometal. Chem. *156*, 307 (1978)
100. Lett, R., Chassaing, G.: Tetrahedron *34*, 2705 (1978)
101. Pearson, D. E., Cowan, D., Beckler, J. D.: J. Org. Chem. *24*, 504 (1959)
102. Luche, J. L., Damiano, J. C.: J. Am. Chem. Soc. *102*, 7926 (1980)
103. Rieke, R. D.: Accts. Chem. Res. *10*, 301 (1977)
104. Lawrence, L. M., Whitesides, G. M.: J. Am. Chem. Soc. *102*, 2493 (1980)
105. Bodewitz, H. W. H. J., Blomberg, C., Bickelhaupt, F.: Tetrahedron *31*, 1053 (1975)
106. Schaart, B. J. et al.: J. Am. Chem. Soc. *98*, 3712 (1976)
107. Wakefield, B. J.: Organometal. Chem. Rev. *1*, 131 (1966)
108. Strohmeier, W., Seifert, F.: Chem. Ber. *94*, 2356 (1961)
109. Houben, J.: Ber. Dtsch. Chem. Ges. *36*, 2897 (1903)
110. Benkeser, R. A.: Synthesis *1971*, 347
111. Gilman, H., McGlumphy, J. H.: Bull. Soc. Chim. France *43*, 1322 (1928)
112. Blomberg, C., Hartog, F. A.: Synthesis *1977*, 18
113. Rieke, R. D., Bales, S. E.: Org. Syn. *59*, 85 (1979)
114. Normant, H., Perrin, P.: Bull. Soc. Chim. France *1957*, 801
115. Freijee, F. et al.: Heterocycles *7*, 237 (1977)
116. Bryce-Smith, D., Skinner, A. C.: J. Chem. Soc. *1963*, 577
117. Gowenlock, B. G., Lindsell, W. E., Singh, B.: J. Chem. Soc. (Dalton) *1978*, 657
118. Esmay, D. L.: Adv. Chem. Ser. *23*, 46 (1959)

119. Gilman, H., Zoellner, E. A., Selby, W. M.: J. Am. Chem. Soc. *55*, 1252 (1933)
120. Kamienski, C. W., Esmay, D. L.: J. Org. Chem. *25*, 1807 (1960)
121. Eisch, J. J., Jacobs, A. M.: J. Org. Chem. *28*, 2145 (1963)
122. Screttas, C. G., Micha-Screttas, M.: J. Org. Chem. *43*, 1064 (1978)
123. Screttas, C. G., Micha-Screttas, M.: J. Org. Chem. *44*, 713 (1979)
124. Cohen, T., Matz, J. R.: J. Am. Chem. Soc. *102*, 6900 (1980)
125. Schlosser, M. in Foerst, W.: Newer Methods of Preparative Organic Chemistry *5*, p. 239, New York: Academic Press 1968
126. Gau, G.: J. Organometal. Chem. *121*, 1 (1976)
127. Morton, A. A., Lanpher, E. J.: J. Org. Chem. *23*, 1636 (1958)
128. Hurd, C. D., Oliver, G. L.: J. Am. Chem. Soc. *81*, 2795 (1959)
129. Rathke, M. W.: Org. React. *22*, 423 (1975)
130. Bellasoued, M., Gaudemar, M.: J. Organometal. Chem. *102*, 1 (1975)
131. Killinger, T. A. et al.: J. Organometal. Chem. *124*, 131 (1977)
132. Rieke, R. D., Uhm, S. J., Hudnall, P. M.: Chem. Commun. *1973*, 269
133. Jones, R. G., Gilman, H.: Org. React. *6*, 339 (1951)
134. Winkler, H. J. S., Winkler, H.: J. Am. Chem. Soc. *88*, 964, 969 (1966)
135. Applequist, D. E., O'Brien, D. F.: J. Am. Chem. Soc. *85*, 743 (1963)
136. Ward, H. R., Lawler, R. G., Cooper, R. A.: J. Am. Chem. Soc. *91*, 746 (1969)
137. Lepley, A. R., Landau, R. L.: J. Am. Chem. Soc. *91*, 748 (1969)
138. Neumann, H., Seebach, D.: Tetrahedron Lett. *1976*, 4839
139. Neumann, H., Seebach, D.: Chem. Ber. *111*, 2785 (1978)
140. Duhamel, L., Poirer, J. M.: J. Am. Chem. Soc. *99*, 8356 (1977)
141. Duhamel, L., Tombret, F.: J. Org. Chem. *46*, 3741 (1981)
142. Hergrueter, C. A. et al.: Tetrahedron Lett. *1977*, 4145, 4573
143. Parham, W. E., Bradsher, C. K., Hunt, D. A.: J. Org. Chem. *43*, 1606 (1978)
144. Akgün, E. et al.: J. Org. Chem. *46*, 2730 (1981)
145. Porzi, G., Concilio, C.: J. Organometal. Chem. *128*, 95 (1977)
146. Newkome, G. R., Roper, J. M.: J. Organometal. Chem. *186*, 147 (1980)
147. Streitwieser, A. Jr., Juaristi, E., Nebenzahl, L. L. in Buncel, E., Durst, T.: Comprehensive Carbanion Chemistry, p. 323, New York: Elsevier 1980
148. Bordwell, F. G. et al.: J. Org. Chem. *45*, 3305 (1980)
149. Bordwell, F. G.: Pure Appl. Chem. *49*, 963 (1977)
150. Bordwell, F. G., Bartmess, J. E., Hautala, J. A.: J. Org. Chem. *43*, 3095 (1978)
151. Bordwell, F. G., Drucker, G. E., Fried, H. E.: J. Org. Chem. *46*, 632 (1981)
152. Olmstead, W. N., Margolin, Z., Bordwell, F. G.: J. Org. Chem. *45*, 3295 (1980)
153. Bordwell, F. G., Drucker, G. E.: J. Org. Chem. *45*, 3325 (1980)
154. Bordwell, F. G., Matthews, W. S., Vanier, N. R.: J. Am. Chem. Soc. *97*, 442 (1975)
155. Bordwell, F. G. et al.: J. Org. Chem. *42*, 321 (1977)
156. Bordwell, F. G., Algrim, D., Fried, H. E.: J. Chem. Soc. (Perkin II) *1979*, 726
157. Bordwell, F. G., Fried, H. E.: J. Org. Chem. *46*, 4327 (1981)
158. Streitwieser, A., Ewing, S. P.: J. Am. Chem. Soc. *97*, 190 (1975)
159. Bordwell, F. G. et al.: J. Org. Chem. *45*, 3884 (1980)
160. Buncel, E., Menon, B.: Chem. Commun. *1976*, 648
161. Algrim, D. et al.: J. Org. Chem. *43*, 5024 (1978)
162. Peterson, D. J.: J. Org. Chem. *32*, 1717 (1967)
163. Bates, R. B., Caldwell, E. S., Klein, H. P.: J. Org. Chem. *34*, 2615 (1969)
164. Olah, G. A., Hunadi, R. J.: J. Org. Chem. *46*, 715 (1981)
165. Ogliaruso, M., Rieke, R., Winstein, S.: J. Am. Chem. Soc. *88*, 4731 (1966)
166. Barfield, M. et al.: J. Am. Chem. Soc. *97*, 900 (1975)
167. Haddon, R. C.: J. Org. Chem. *44*, 3608 (1979)

168. Märkl, G., Liebl, R.: Ann. Chem. *1980*, 2095
169. Maercker, A.: Org. React. *14*, 270 (1965)
170. Lowe, P. A.: Chem. & Ind. *1970*, 1070
171. Gschwend, H. W., Rodriguez, H. R.: Org. React. *26*, 1 (1979)
172. Peterson, D. J.: J. Organometal. Chem. *9*, 373 (1967)
173. Magnus, P.: Aldrichimica Acta *13*, 43 (1980)
174. Meyers, A. I., Ten Hoeve, W.: J. Am. Chem. Soc. *102*, 7125 (1980)
175. Seebach, D., Hassel, T.: Angew. Chem., Int. Ed. Engl. *90*, 274 (1978)
176. Beak, P., Reitz, D. B.: Chem. Rev. *78*, 275 (1978)
177. Magnus, P., Roy, G.: Synthesis *1980*, 575
178. Schöllkopf, U.: Angew. Chem., Int. Ed. Engl. *16*, 339 (1977)
179. Schöllkopf, U.: Pure and Appl. Chem. *51*, 1347 (1979)
180. Peterson, D. J.: J. Organometal. Chem. *8*, 199 (1967)
181. Mathey, F., Mercier, F.: J. Organometal. Chem. *177*, 255 (1979)
182. Beak, P., Baillargeon, M., Carter, L. G.: J. Org. Chem. *43*, 4255 (1978)
183. Beak, P., Carter, L. G.: J. Org. Chem. *46*, 2363 (1981)
184. Kutsuma, T. et al.: Heterocycles *8*, 397 (1977)
185. Field, L.: Synthesis *1978*, 713
186. Hayashi, T., Sakurai, A., Oishi, T.: Chem. Lett. *1977*, 1483
187. Tamura, Y. et al.: Synthesis *1977*, 693
188. Welch, S. C., Rao, A. S. C. P.: J. Am. Chem. Soc. *101*, 6135 (1979)
189. Dumont, W., Sevrin, M., Krief, A.: Angew. Chem., Int. Ed. Engl. *16*, 541 (1977)
190. Sevrin, M., Krief, A.: Tetrahedron Lett. *1978*, 187
191. Göbel, B. T., Seebach, D.: Angew. Chem., Int. Ed. Engl. *13*, 83 (1974)
192. Magnus, P., Roy, G.: Chem. Commun. *1979*, 822
193. Ager, D. J., Cookson, R. C.: Tetrahedron Lett. *21*, 1677 (1980)
194. Sachdev, K., Sachdev, H. S.: Tetrahedron Lett. *1976*, 4223
195. Burford, C. et al.: J. Am. Chem. Soc. *99*, 4536 (1977)
196. Cooke, F., Magnus, P.: Chem. Commun. *1977*, 513
197. Broekhof, N. L. J. M., Jonkers, F. L., van der Gen, A.: Tetrahedron Lett *1979*, 2433
198. Earnshaw, C., Wallis, C. J., Warren, S.: Chem. Commun. *1977*, 314
199. Grayson, J. I., Warren, S.: J. Chem. Soc. (Perkin I) *1977*, 2263
200. Savignac, P., Coutrot, P.: Synthesis *1978*, 682
201. Yamashita, M., Suemitsu, R.: Chem. Commun. *1977*, 691
202. Neube, S. et al.: Tetrahedron Lett. *1978*, 2345
203. Ogura, K. et al.: Synthesis *1979*, 880
204. Reutrakul, V., Kanghae, W.: Tetrahedron Lett. *1977*, 1225
205. Taguchi, H., Yamamoto, H., Nozaki, H.: Bull. Chem. Soc. Japan *50*, 159? (1977)
206. Cohen, T., Nolan, S. M.: Tetrahedron Lett. *1978*, 3533
207. Compere, E. L., Shockravi, A.: J. Org. Chem. *43*, 2702 (1978)
208. Dessy, R. E. et al.: J. Am. Chem. Soc. *88*, 460 (1966)
209. Klumpp, G. W. et al.: J. Am. Chem. Soc. *101*, 7065 (1979)
210. Bates, R. B., Beavers, W. A., Blacksberg, I. R.: Abstracts, 169th National AC? Meeting, April, 1975, ORGN-16
211. Schmidt, R. R., Talbiersky, J., Russegger, P.: Tetrahedron Lett. *1979*, 4273
212. Schöllkopf, U., Stafforst, D., Jentsch, R.: Ann. Chem. *1977*, 1167
213. Oakes, F. T., Sebastian, J. F.: J. Org. Chem. *45*, 4959 (1980)
214. Marino, J. P., Katterman, L. C.: Chem. Commun. *1979*, 946
215. Knight, D. W.: Tetrahedron Lett. *1979*, 469

216. Corey, E. J., Boger, D. L.: Tetrahedron Lett. *1978*, 5
217. Chadwick, D. J., Willbe, C.: J. Chem. Soc. (Perkin I) 887 (1977)
218. Slocum, D. W., Sugarman, D. I.: Adv. Chem. Ser. *130*, 222 (1974)
219. Beak, P., Brown, R. A.: J. Org. Chem. *44*, 4463 (1979)
220. Meyers, A. I., Lutomski, K.: J. Org. Chem. *44*, 4464 (1979)
221. Ronald, R. C.: Tetrahedron Lett. *1975*, 3973
222. Walborsky, H. M., Ronman, P.: J. Org. Chem. *43*, 731 (1978)
223. Omae, I.: Chem. Rev. *79*, 287 (1979)
224. Barsky, L. et al.: J. Org. Chem. *41*, 3651 (1976)
225. Harris, T. D., Roth, G. P.: J. Org. Chem. *44*, 2004 (1979)
226. Bruce, M. I.: Angew. Chem., Int. Ed. Engl. *16*, 73 (1977)
227. Katritsky, A. R., Rahimi-Rastgoo, S., Ponkshe, N. K.: Synthesis *1981*, 127
228. Meyer, N., Seebach, D.: Angew. Chem., Int. Ed. Engl. *17*, 521 (1978)
229. Meyer, N., Seebach, D.: Chem. Ber. *113*, 1304 (1980)
230. Panetta, C. A., Dixit, A. S.: Synthesis *1981*, 59
231. Meyers, A. I., Gabel, R. A.: Tetrahedron Lett. *1978*, 227
232. Puterbaugh, W. H., Hauser, C. R.: J. Am. Chem. Soc. *85*, 2467 (1963)
233. Beak, P., Brown, R. A.: J. Org. Chem. *42*, 1823 (1977)
234. Baldwin, J. E., Blair, K. W.: Tetrahedron Lett. *1978*, 2559
235. Jones, J. R.: Surv. Prog. Chem. *6*, 83 (1973)
236. House, H. O.: Rec. Chem. Prog. *28*, 99 (1967)
237. Fleming, I., Paterson, I.: Synthesis *1979*, 736
238. Pellerite, M. J., Brauman, J. I. in Buncel, E., Durst, T.: Comprehensive Carbanion Chemistry, p. 55, New York: Elsevier (1980)
239. Stevenson, G. R., Williams, E., Jr.: J. Am. Chem. Soc. *101*, 5910 (1979)
240. Bordwell, F. G. et al.: J. Am. Chem. Soc. *97*, 3226 (1975)
241. Cumming, J. B., Kebarle, P.: Can. J. Chem. *56*, 1 (1978)
242. Bartmess, J. E., Scott, J. A., McIver, R. T., Jr.: J. Am. Chem. Soc. *101*, 6046 (1979)
243. Bartmess, J. E., McIver, R. T., Jr. in Bowers, M. T.: Gas-Phase Ion Chemistry, New York: Academic Press 1978
244. Mackay, G. I., Hemsworth, R. S., Bohme, D. K.: Can. J. Chem. *54*, 1624 (1976)
245. Dewar, M. J. S., Fox, M. A., Nelson, D. J.: J. Organometal. Chem. *185*, 157 (1980)
246. Ilić, P., Trinajstić, N.: J. Org. Chem. *45*, 1738 (1980)
247. Pross, A. et al.: J. Org. Chem. *46*, 1693 (1981)
248. Bates, R. B. et al.: J. Am. Chem. Soc. *103*, 5052 (1981)
249. Bank, S.: J. Org. Chem. *37*, 114 (1972)
250. Steiner, E. C., Gilbert, J. M.: J. Am. Chem. Soc. *85*, 3054 (1963)
251. Nielsen, A. T., Houlihan, W. J.: Org. React. *16*, 1 (1968)
252. Gokel, G. W., Weber, W. P.: J. Chem. Ed. *55*, 350 (1978)
253. Sjöberg, K.: Aldrichimica Acta *13*, 55 (1980)
254. Rebuffat, S., Giraud, M., Molho, D.: Bull. Soc. Chim. France *1978*, 457
255. Golinski, J., Makosza, M.: Synthesis *1978*, 823
256. Pearson, D. E., Buehler, C. A.: Chem. Rev. *74*, 45 (1974)
257. Orr, D.: Synthesis *1979*, 139
258. Ono, N. et al.: Bull. Chem. Soc. Japan *52*, 1716 (1979)
259. Harris, T. M., Harris, C. M.: Org. React. *17*, 155 (1969)
260. Kaiser, E. M., Petty, J. D., Knutson, P. L. A.: Synthesis *1977*, 509
261. Brown, C. A., Yamashita, A.: J. Am. Chem. Soc. *97*, 891 (1975)
262. Shatenshtein, A. I., Shapiro, I. O.: Russ. Chem. Rev. *37*, 845 (1968)

XI. References

263. Olofson, R. A., Dougherty, C. M.: J. Am. Chem. Soc. *95*, 581, 582 (1973)
264. Gygax, P., Eschenmoser, A.: Helv. Chim. Acta *60*, 507 (1977)
265. Schaefer, J. P., Bloomfield, J. J.: Org. React. *15*, 1 (1967)
266. Brown, C. A.: J. Org. Chem. *39*, 3913 (1974)
267. Corey, E. J., Chaykovsky, M.: J. Am. Chem. Soc. *84*, 866 (1962)
268. Ziegenbein, W. in Viehe, H. G.: Chemistry of Acetylenes, p. 169, New York: Marcel Dekker, 1969
269. Savoia, D., Trombini, C., Umani-Ronchi, A.: Tetrahedron Lett. *1977*, 653
270. Hart, H., Chen, B., Peng, C.: Tetrahedron Lett. *1977*, 3121
271. Savoia, D. et al.: J. Organometal. Chem. *204*, 281 (1981)
272. Gilman, H., Morton, J. W.: Org. React. *8*, 258 (1954)
273. Benkeser, R. A., Foster, D. J., Sauve, D. M.: Chem. Rev. *57*, 867 (1957)
274. Mallan, J. M., Bebb, R. L.: Chem. Rev. *69*, 693 (1969)
275. Bates, R. B., Gosselink, D. W., Kaczynski, J. A.: Tetrahedron Lett. *1967*, 199
276. Eberhardt, G. G., Botte, W. A.: J. Org. Chem. *29*, 2928 (1964)
277. Agami, C.: Bull. Soc. Chim. France *1970*, 1619
278. Trimitsis, G. B. et al.: J. Org. Chem. *38*, 1491 (1973)
279. Akiyama, S., Hooz, J.: Tetrahedron Lett. *1973*, 4115
280. Klein, J., Medlik, A.: Chem. Commun. *1973*, 275
281. Crawford, R. J., Erman, W. F., Broaddus, C. D.: J. Am. Chem. Soc. *94*, 4298 (1972)
282. Broaddus, C. D.: J. Org. Chem. *29*, 2689 (1964)
283. Crimmins, T. F., Rather, E. M.: J. Org. Chem. *43*, 2170 (1978)
284. Lochmann, L., Pospíšil, J., Lím, D.: Tetrahedron Lett. *1966*, 257
285. Schlosser, M.: J. Organometal. Chem. *8*, 9 (1967)
286. Benkeser, R. A., Crimmons, T. F., Tong, W.: J. Am. Chem. Soc. *90*, 4366 (1968)
287. Hartmann, J., Schlosser, M.: Helv. Chem. Acta *59*, 453 (1976)
288. Bahl, J. J., Bates, R. B., Gordon, B.: J. Org. Chem. *44*, 2290 (1979)
289. Schlosser, M., Hartmann, J.: J. Am. Chem. Soc. *98*, 4674 (1976)
290. Hart, A. J., O'Brien, D. H., Russell, C. R.: J. Organometal. Chem. *72*, C19 (1974)
291. Burwell, R. L.: Chem. Rev. *54*, 615 (1954)
292. Bowers, K. W. et al.: J. Am. Chem. Soc. *92*, 2783 (1970)
293. Bates, R. B., Kroposki, L. M., Potter, D. E.: J. Org. Chem. *37*, 560 (1972)
294. Lipton, M. F. et al.: J. Organometal. Chem. *186*, 155 (1980)
295. Bergbreiter, D. E., Pendergrass, E.: J. Org. Chem. *46*, 219 (1981)
296. Katz, T. J.: J. Am. Chem. Soc. *82*, 3784 (1960)
297. Cox, R. H., Harrison, L. W., Austin, W. K., Jr.: J. Phys. Chem. *77*, 200 (1973)
298. Holy, N. L.: Chem. Rev. *74*, 243 (1974)
299. Yang, M. et al.: Tetrahedron Lett. *1970*, 3843
300. Baker, R., Cookson, R. C., Saunders, A. D.: J. Chem. Soc. (Perkin I) *1976*, 1809, 1815
301. Akutagawa, S., Otsuka, S.: J. Am. Chem. Soc. *98*, 7420 (1976)
302. Kajihara, Y. et al.: Bull. Chem. Soc. Japan *53*, 3035 (1980)
303. Yasuda, H. et al.: Bull. Chem. Soc. Japan *52*, 2036 (1979)
304. Engels, R., Schäfer, H. J.: Angew. Chem., Int. Ed. Engl. *17*, 460 (1978)
305. Shono, T., Nishiguchi, I., Ohmizu, H.: J. Am. Chem. Soc. *99*, 7396 (1977)
306. Lund, H., Degrand, C.: Tetrahedron Lett. *1977*, 3593
307. Bergman, E. D., Ginsburg, D., Pappo, R.: Org. React. *10*, 179 (1959)
308. Strauss, M. J.: Chem. Rev. *70*, 667 (1970)
309. Leffler, M. T.: Org. React. *1*, 91 (1942)

310. Wender, P. A., Eissenstat, M. A.: J. Am. Chem. Soc. *100*, 292 (1978)
311. Wender, P. A., Schaus, J. M.: J. Org. Chem. *43*, 782 (1978)
312. Martin, S. F. et al.: J. Am. Chem. Soc. *102*, 5866 (1980)
313. Fraenkel, G., Estes, D., Geckle, M. J.: J. Organometal. Chem. *185*, 147 (1980)
314. Bates, R. B. et al.: J. Am. Chem. Soc. *91*, 4608 (1969)
315. Szwarc, M.: Carbanions Living Polymers and Electron Transfer Processes, New York: Wiley 1968
316. Posner, G. H., Lentz, C. M.: Tetrahedron Lett. *1977*, 3215
317. Heng, K. K., Smith, R. A. J.: Tetrahedron *35*, 425 (1979)
318. Coates, G. E., Wade, K.: Organometallic Compounds, *1*, The Main Group Elements, London: Methuen 1967
319. Makarova, L. G., Nesmeyanov, A. N.: Methods of Elemento-Organic Chemistry *4*, Amsterdam: North-Holland Publishing 1967
320. Fraenkel, G., Dix, D. T., Carlson, M.: Tetrahedron Lett. *1968*, 579
321. O'Brien, D. H., Russell, C. R., Hart, A. J.: J. Am. Chem. Soc. *101*, 633 (1979)
322. Schlenk, W., Holtz, J.: Ber. Dtsch. Chem. Ges. *50*, 262 (1917)
323. Wittig, G., Bickelhaupt, F.: Chem. Ber. *91*, 883 (1958)
324. Barluenga, J., Fañanás, F., Yus, M.: J. Org. Chem. *44*, 4798 (1979)
325. Barluenga, J., Fañanás, F. J., Yus, M.: J. Org. Chem. *46*, 1281 (1981)
326. Seyferth, D.: Rec. Chem. Progr. *26*, 87 (1965)
327. Seyferth, D., Weiner, M. A.: J. Am. Chem. Soc. *83*, 3583 (1961)
328. Seyferth, D., Weiner, M. A.: J. Org. Chem. *26*, 4797 (1961)
329. Seyferth, D., Jula, T. F.: J. Organometal. Chem. *66*, 195 (1974)
330. Seyferth, D., Mammarella, R. E.: J. Organometal. Chem. *177*, 53 (1979)
331. Märkl, G., Liebl, R.: Ann. *1980*, 2095
332. Wollenberg, R. H., Albizati, K. F., Peries, R.: J. Am. Chem. Soc. *99*, 7365 (1977)
333. Wollenberg, R. H.: Tetrahedron Lett. *1978*, 717
334. Chen, S. L., Schaub, R. E., Grudzinskas, C. V.: J. Org. Chem. *43*, 3450 (1978)
335. Piers, E., Morton, H. E.: J. Org. Chem. *44*, 3437 (1979)
336. Stork, G., Hudrlik, P. F.: J. Am. Chem. Soc. *90*, 4462, 4464 (1968)
337. Hosomi, A., Shirahata, A., Sakurai, H.: Tetrahedron Lett. *1978*, 3043
338. Sarkar, T. K., Andersen, N. H.: Tetrahedron Lett. *1978*, 3513
339. Trost, B. M., Vincent, J. E.: J. Am. Chem. Soc. *102*, 5680 (1980)
340. Rauchschwalbe, G., Schlosser, M.: Helv. Chim. Acta *58*, 1094 (1975)
341. Marr, G., Moore, R. E., Rockett, B. W.: J. Chem. Soc. (C) *1968*, 24
342. Matteson, D. S., Majumdar, D.: J. Organometal. Chem. *184*, C41 (1980)
343. Levy, A. B. et al.: J. Organometal. Chem. *156*, 123 (1978)
344. Posner, G. H.: Org. React. *22*, 253 (1975)
345. Bertz, S. H., Dabbagh, G.: Abstracts, 183rd National ACS Meeting, March, 1982, ORGN-110
346. Bordwell, F. G., Clemens, A. H.: J. Org. Chem. *46*, 1035 (1981)
347. Normant, J. F., Villieras, J., Scott, F.: Tetrahedron Lett. *1977*, 3263
348. Mandai, T. et al.: Tetrahedron *35*, 309 (1979)
349. Bates, R. B., White, J. J.: Unpublished results
349a. Bates, R. B. et al.: J. Am. Chem. Soc. *82*, 2327 (1960)
349b. Alder, K., Ache, H. J.: Chem. Ber. *95*, 503 (1962)
349c. Deuchert, K., Hertenstein, U., Hünig, S.: Synthesis *1973*, 777
349d. Kristensen, L. H., Lund, H.: Acta Chem. Scand. *B33*, 735 (1979)
349e. Ashe, A. J., Diephouse, T. R., El-Sheikh, M. Y.: J. Am. Chem. Soc. *104*, 5693 (1982)
350. Pettit, G. R., van Tamelen, E. E.: Org. React. *12*, 356 (1962)

351. Julia, M., Blasioli, C.: Bull. Soc. Chim. France *1976*, 1941
352. Meyers, A. I., Tait, T. A., Comins, D. L.: Tetrahedron Lett. *1978*, 4657
353. Hojo, M. et al.: Tetrahedron Lett. *1977*, 3883
354. Petragnani, N., Ferraz, H. M. C.: Synthesis *1978*, 476
355. Maruyama, K., Yamamoto, Y.: J. Am. Chem. Soc. *99*, 8068 (1977)
356. Mukaiyama, T., Imaoka, M., Izawa, T.: Chem. Lett. *1977*, 1257
357. Mukaiyama, T., Yamaguchi, M., Narasaka, K.: Chem. Lett. *1978*, 689
358. Goering, H. L., Kantner, S. S.: J. Org. Chem. *46*, 2144 (1981)
359. Calò, V. et al.: Synthesis *1979*, 885
360. Tadema, G. et al.: Recl. Trav. Chim. *95*, 66 (1976)
361. Sato, F., Kodama, H., Sato, M.: Chem. Lett. *1978*, 789
362. Dreger, E. E.: Org. Syn. Coll. Vol. *1*, 306 (1932)
363. Acker, R. D.: Tetrahedron Lett. *1977*, 3407
364. Schollkopf, V., Jentsch, R., Madawinata, K.: Ann. Chem. *1979*, 451
365. Piers, E., Nagakura, I.: Tetrahedron Lett. *1976*, 3237
366. Commercon, A., Normant, J. F., Villieras, J.: J. Organometal. Chem. *128*, (1977)
367. Hayashi, T. et al.: Tetrahedron Lett. *1981*, 137
368. Dang, H. P., Linstrumelle, G.: Tetrahedron Lett. *1978*, 191
369. Murahashi, S. et al.: J. Org. Chem. *44*, 2408 (1979)
370. King, A. O. et al.: J. Org. Chem. *43*, 358 (1978)
371. Whitesides, G. M. et al.: J. Am. Chem. Soc. *91*, 4871 (1969)
372. Lebouc, A., Delaunay, J., Riobé, O.: Synthesis *1979*, 610
373. Bare, T. M., House, H. O.: Org. Syn. Coll. Vol. *5*, 775 (1973)
374. Newman, M. S., Smith, A. S.: J. Org. Chem. *13*, 592 (1948)
375. Mukaiyama, T., Araki, M., Takei, H.: J. Am. Chem. Soc. *95*, 4763 (1973)
376. Comins, D., Meyers, A. I.: Synthesis *1978*, 403
377. Huet, F., Emptoz, G., Jubier, A.: Tetrahedron *29*, 479 (1973)
378. Scilly, N. F.: Synthesis *1973*, 160
379. Nelson, D. J., Uschak, E. A.: J. Org. Chem. *42*, 3308 (1977)
380. Patel, K. M. et al.: Tetrahedron Lett. *1976*, 4015
381. Cason, J., Prout, F. S.: Org. Syn. Coll. Vol. *3*, 601 (1955)
382. Cahiez, G. et al.: Tetrahedron Lett. *1976*, 3155
383. Pittman, C. U., Hanes, R. M.: J. Org. Chem. *42*, 1194 (1977)
384. Demuth, M.: Helv. Chim. Acta *61*, 3136 (1978)
385. Kahne, D., Collum, D. B.: Tetrahedron Lett. *22*, 5011 (1981)
386. Klein, J., Medlik, A., Meyer, A. Y.: Tetrahedron *32*, 51 (1976)
387. Bates, R. B., Ogle, C. A.: J. Org. Chem. *47*, in press (1982)
388. Boche, G. et al.: Angew. Chem., Int. Ed. Engl. *17*, 687 (1978)
389. Gierer, P.: Ph. D. Thesis, Southern Illinois University, 1972
390. Marr, G., Hunt, T.: J. Chem. Soc. (C) *1969*, 1070
391. Gräfing, R., Brandsma, L.: Synthesis *1978*, 578
392. Nakai, T., Mimura, T.: Tetrahedron Lett. *1979*, 531
393. Dinizo, S. E. et al.: J. Org. Chem. *41*, 2846 (1976)
394. Verboom, W., Meijer, J., Brandsma, L.: Synthesis *1978*, 577
395. Baarschers, W. H.: Can. J. Chem. *54*, 3056 (1976)
396. Adcock, J. L., Renk, E. B.: J. Org. Chem. *44*, 3431 (1979)
397. Harnisch, J. et al.: J. Am. Chem. Soc. *101*, 3370 (1979)
398. Levy, A. B., Talley, P., Dunford, J. A.: Tetrahedron Lett. *1977*, 3545
399. Gay, R. L., Crimmins, T. F., Hauser, C. R.: Chem. & Ind. *1966*, 1635
400. Arnold, R. T., Kulenovic, S. T.: J. Org. Chem. *43*, 3687 (1978)

401. Cotton, F. A., Wilkinson, G.: Advanced Inorganic Chemistry, 4th Ed., New York: Wiley 1980
402. Yamashita, M., Suemitsu, R.: Tetrahedron Lett. 1978, 761
403. Kamienski, C. W.: Adv. Chem. Ser. *130*, 163 (1974)
404. Ager, D. J.: Tetrahedron Lett. *22*, 587 (1981)
405. Kauffmann, T. et al.: Angew. Chem., Int. Ed. Engl. *16*, 710 (1977)
406. Bunting, W., Langer, A. W.: Adv. Chem. Ser. *130*, 201 (1974)
407. Bahl, J. J. et al.: J. Am. Chem. Soc. *99*, 6126 (1977)
408. Hallden-Abberton, M., Engelman, C., Fraenkel, G.: J. Org. Chem. *46*, 538 (1981)
409. Gausing, W. et al.: J. Organometal. Chem. *199*, 137 (1980)
410. Posner, G. H.: Org. React. *19*, 1 (1972)
411. Liu, S. H.: J. Org. Chem. *42*, 3209 (1977)
412. Suzuki, M. et al.: Tetrahedron Lett. *21*, 1247 (1980)
413. Ostrowski, P. C., Kane, V. V.: Tetrahedron Lett. *1977*, 3549
414. Brown, C. A., Yamaichi, A.: Chem. Commun. *1979*, 100
415. Clive, D. L. J., Farina, V., Beaulieu, P. L.: J. Org. Chem. *47*, 2572 (1982)
416. Isobe, M. et al.: Chem. Lett. *1977*, 679
417. Seebach, D., Locher, R.: Angew. Chem., Int. Ed. Engl. *18*, 957 (1979)
418. Corey, E. J., Boger, D. L.: Tetrahedron Lett. *1978*, 5, 9, 13
419. Meyers, A. I., Smith, R. K., Whitten, C. E.: J. Org. Chem. *44*, 2250 (1979)
420. Bruson, H. A.: Org. React. *5*, 79 (1949)
421. Bartoli, G. et al.: Synthesis *1978*, 436
422. Marfat, A., McGuirk, P. R., Helquist, P.: Tetrahedron Lett. *1978*, 1363
423. Smorada, R. L., Truce, W. E.: J. Org. Chem. *44*, 3444 (1979)
424. Harada, K. in Patai, S.: Chemistry of the Carbon-Nitrogen Double Bond, p. 266, New York: Interscience 1970
425. Scully, F. E.: J. Org. Chem. *45*, 1515 (1980)
426. Davis, F. A., Mancinelli, P. A.: J. Org. Chem. *42*, 398 (1977)
427. Giam, C. S. et al.: J. Chem. Soc. (Perkin I) *1979*, 3082
428. Roberts, J. L., Borromeo, P. S., Poulter, C. D.: Tetrahedron Lett. *1977*, 1299
429. Eisele, G., Simchen, G.: Synthesis *1978*, 757
430. Kharasch, M. S., Reinmuth, O.: Grignard Reactions of Nonmetallic Substances, p. 767, Englewood Cliffs, N.J.: Prentice-Hall 1954
431. Gauthier, R., Chastrette, M.: J. Organometal. Chem. *165*, 139 (1979)
432. Eicher, T. in Patai, S.: Chemistry of the Carbonyl Group, p. 621, New York: Interscience 1966
433. Buhler, J. D.: J. Org. Chem. *38*, 904 (1973)
434. Okubo, M., Morigami, Y., Suenaga, R.: Bull. Chem. Soc. Japan *53*, 3029 (1980)
435. Ashby, E. C., Goel, A. B.: J. Am. Chem. Soc. *103*, 4983 (1981)
436. Mazaleyat, J.-P., Cram, D. J.: J. Am. Chem. Soc. *103*, 4585 (1981)
437. Molle, G., Bauer, P.: J. Am. Chem. Soc. *104*, 3481 (1982)
438. Bestmann, H. J., Zimmermann, R. in Augustine, R. L.: Carbon-Carbon Bond Formation, p. 353, New York: Marcel Dekker 1979
439. Boutagy, J., Thomas, R.: Chem. Rev. *74*, 87 (1974)
440. Loupy, A., Sogadji, K., Seyden-Penne, J.: Synthesis *1977*, 126
441. Coutrot, P., Snoussi, M., Savignac, P.: Synthesis *1978*, 133
442. Brittelli, D. R.: J. Org. Chem. *46*, 2514 (1981)
443. Peterson, D.: J. Org. Chem. *33*, 780 (1968)
444. Hartzell, S. L., Rathke, M. W.: Tetrahedron Lett. *1976*, 2737
445. Konakahara, T., Takagi, Y.: Synthesis *1979*, 192

446. Ueno, Y., Setoi, M., Okawara, M.: Chem. Lett. *1979*, 47
447. Kauffmann, T., Kriegesmann, R., Woltermann, A.: Angew. Chem., Int. Ed Engl. *16*, 862 (1977)
448. Tanaka, K. et al.: Chem. Lett. *1978*, 197
449. Tanaka, K., Tanikaga, R., Kaji, A.: Chem. Lett. *1976*, 917
450. Volpin, M. E., Kolomnikov, I. S.: Organometal. React. *5*, 313 (1975)
451. Ryang, M. et al.: Bull. Chem. Soc. Japan *37*, 1704 (1964)
452. Trzupek, L. S. et al.: J. Am. Chem. Soc. *95*, 8118 (1973)
453. Baird, D. M., Bereman, R. D.: J. Org. Chem. *46*, 458 (1981)
454. Wollenberg, R. H., Albizati, K. F., Peries, R.: J. Am. Chem. Soc. *99*, 7365 (1977)
455. Chandrasekhar, J., Andrade, J. G., Schleyer, P. v. R.: J. Am. Chem. Soc. *103* 5612 (1981)
456. Normant, J. F. et al.: Tetrahedron Lett. *1978*, 3711
457. Westmijze, H., Meijer, J., Vermeer, P.: Recl. Trav. Chim. *96*, 168 (1977)
458. Bradsher, C. K., Reames, D. C.: J. Org. Chem. *43*, 3800 (1978)
459. Bodem, G. B., Leete, E.: J. Org. Chem. *44*, 4696 (1979)
460. Midland, M. M.: J. Org. Chem. *42*, 2650 (1977)
461. Verboom, W. et al.: Synthesis *1979*, 296
462. Hiyama, T., Shinoda, M., Nozaki, H.: J. Am. Chem. Soc. *101*, 1599 (1979)
463. Hauptmann, H., Mader, M.: Synthesis *1978*, 307
464. Ng, K. S., Alper, H.: J. Organometal. Chem. *202*, 1 (1980)
465. Carlson, R. M.: Tetrahedron Lett. *1978*, 111
466. Wilson, D. R., DiLullo, A. A., Ernst, R. D.: J. Am. Chem. Soc. *102*, 5928 (1980)
467. Burger, U., Gandillon, G.: Tetrahedron Lett. *1979*, 4281
468. Volger, H. C., Brackman, W.: Recl. Trav. Chim. *84*, 1017 (1965)
469. de Graff, S. A. G., Oosterhoff, P. E. R., van der Gen, A.: Tetrahedron Lett *1974*, 1653
470. Katzenellenbogen, J. A., Crumrine, A. L.: J. Am. Chem. Soc. *98*, 4925 (1976)
471. Ingold, C. K., de Salas, E., Wilson, C. L.: J. Chem. Soc. *1936*, 1328
472. Fraser, H. B., Kon, G. A. R.: J. Chem. Soc. *1934*, 604
473. Nakai, T., Mimura, T., Ari-Izumi, A.: Tetrahedron Lett. *1977*, 2425
474. Broaddus, C. D.: Accts. Chem. Res. *1*, 231 (1968)
475. Savoia, D., Trombini, C., Umani-Ronchi, A.: J. Chem. Soc. (Perkin I) *1977* 123
476. Macdonald, T. L., Narayanan, B. A., O'Dell, D. E.: J. Org. Chem. *46*, 1504 (1981)
477. Hiyama, T., Shinoda, M., Nozaki, H.: Tetrahedron Lett. *1978*, 771
478. Bates, R. B. et al.: J. Am. Chem. Soc. *95*, 926 (1973)
479. Bosshardt, H., Schlosser, M.: Helv. Chim. Acta *63*, 2393 (1980)
480. Ela, S. W., Cram, D. J.: J. Am. Chem. Soc. *88*, 5791 (1966)
481. Ahlbrecht, H. et al.: Tetrahedron Lett. *21*, 3175 (1980)
482. Evans, D. A. et al.: J. Am. Chem. Soc. *102*, 5955 (1980)
483. Bates, R. B., Beavers, W. A.: J. Am. Chem. Soc. *96*, 5001 (1974)
484. Linstrumelle, G., Lorne, R., Dang, H. P.: Tetrahedron Lett. *1978*, 4069
485. Majewski, M. et al.: J. Org. Chem. *46*, 2029 (1981)
486. Babler, J. H., Invergo, B. J.: J. Org. Chem. *46*, 1937 (1981)
487. Krull, I. S., Arnold, D. R.: Org. Prep. Proc. *1*, 283 (1969)
488. Barbot, F., Miginiac, P.: Bull. Soc. Chim. France *1977*, 113
489. Reich, H. J., Shah, S. K.: J. Am. Chem. Soc. *99*, 263 (1977)
490. Reich, H. J. et al.: J. Am. Chem. Soc. *103*, 3112 (1981)

491. House, H. O. et al.: J. Org. Chem. *43*, 700 (1978)
492. Olofson, R. A., Cuomo, J., Bauman, B. A.: J. Org. Chem. *43*, 2073 (1978)
493. Larson, G. L., Fuentes, L. M.: J. Am. Chem. Soc. *103*, 2418 (1981)
494. Tardella, P. A.: Tetrahedron Lett. *1969*, 1117
495. Rathke, M. W., Lindert, A.: Synth. Commun. *8*, 9 (1978)
496. Groenewegen, P., Kallenberg, H., van der Gen, A.: Tetrahedron Lett. *1978*, 491
497. Dietl, H. K., Brannock, K. C.: Tetrahedron Lett. *1973*, 1273
498. Kuehne, M. E.: Synthesis *1970*, 510
499. Caine, D. in Augustine, R. L.: Carbon-Carbon Bond Formation, p. 85, New York: Marcel Dekker 1979
500. Piers, E., Abeysekera, B., Scheffer, J. R.: Tetrahedron Lett. *1979*, 3279
501. Millard, A. A., Rathke, M. W.: J. Am. Chem. Soc. *99*, 4833 (1977)
502. Beugelmans, R., Roussi, G.: Chem. Commun. *1979*, 950
503. Bard, R. R., Bunnett, J. F.: J. Org. Chem. *45*, 1546 (1980)
504. Semmelhack, M. F. et al.: J. Am. Chem. Soc. *97*, 2507 (1975)
505. Bunnett, J. F.: Accts. Chem. Res. *11*, 413 (1978)
506. Moon, M. P., Wolfe, J. F.: J. Org. Chem. *44*, 4081 (1979)
507. Semmelhack, M. F., Bargar, T.: J. Am. Chem. Soc. *102*, 7765 (1980)
508. Russell, G. A., Mudryk, B., Jawdosiuk, M.: J. Am. Chem. Soc. *103*, 4610 (1981)
509. Beck, A. K., Hoekstra, M. S., Seebach, D.: Tetrahedron Lett. *1977*, 1187
510. Howard, A. S., Meerholz, C. A., Michael, J. P.: Tetrahedron Lett. *1979*, 1339
511. Montforts, F. P., Ofner, S.: Angew. Chem., Int. Ed. Engl. *18*, 632 (1979)
512. Hauser, C. R., Hudson, B. E., Jr.: Org. React. *1*, 266 (1942)
513. Bartmess, J. E., Hays, R. L., Caldwell, G.: J. Am. Chem. Soc. *103*, 1338 (1981)
514. Couffignal, R., Moreau, J. L.: Tetrahedron Lett. *1978*, 3713
515. Blaszczak, L., Winkler, J., O'Kuhn, S.: Tetrahedron Lett. *1976*, 4405
516. Hajos, Z. G. in Augustine, R. L.: Carbon-Carbon Bond Formation, p. 1, New York: Marcel Dekker 1979
517. Noyori, R., Nishida, I., Sakata, J.: J. Am. Chem. Soc. *103*, 2106 (1981)
518. Kraus, G. A., Taschner, M. J.: Tetrahedron Lett. *1977*, 4575
519. Wittig, G., Hesse, A.: Org. Syn. *50*, 66 (1970)
520. Miller, J. A., Zweifel, G.: J. Am. Chem. Soc. *103*, 6217 (1981)
521. Kotake, H., Inomata, K., Sumita, M.: Chem. Lett. *1978*, 717
522. Ballester, M.: Chem. Rev. *55*, 283 (1955)
523. Bartlett, P. A.: Tetrahedron *36*, 3 (1980)
524. Heathcock, C. H.: Science *214*, 395 (1981)
525. Masamune, S. et al.: J. Am. Chem. Soc. *103*, 1566, 1568 (1981)
526. House, H. O. et al.: J. Am. Chem. Soc. *95*, 3310 (1973)
527. Evans, D. A., McGee, L. R.: J. Am. Chem. Soc. *103*, 2876 (1981)
528. Jeffrey, E. A., Meisters, A., Mole, T.: J. Organometal. Chem. *74*, 373 (1974)
529. Yamamoto, Y. et al.: J. Am. Chem. Soc. *102*, 7107 (1980)
530. Wartski, L.: Chem. Commun. *1977*, 602
531. Roberts, J. L., Borromeo, P. S., Poulter, C. D.: Tetrahedron Lett. *1977*, 1621
532. Garst, M. E.: J. Org. Chem. *44*, 1578 (1979)
533. Katritzky, A. R. et al.: Angew. Chem., Int. Ed. Engl. *18*, 792 (1979)
534. Carre, M. C., Viriot-Villaume, M. L., Caubere, P.: Synthesis *1977*, 48
535. Tramontini, M.: Synthesis *1973*, 703
536. Krause, G. A., Sugimoto, H.: Tetrahedron Lett. *1977*, 3929
537. Jung, M. E., McCombs, C. A.: Tetrahedron Lett. *1976*, 2935
538. Kende, A. S. et al.: Tetrahedron Lett. *1979*, 4513
539. Hagiwara, H. et al.: Chem. Commun. *1976*, 413

540. Mander, L. N., Woolias, M.: Synthesis *1979*, 185
541. Stetter, H. in Houben-Weyl: Methoden der Organischen Chemie *7/2b*, p. 1423 Stuttgart: Thieme 1976
542. Henecka, H. in Houben-Weyl: Methoden der Organischen Chemie *7/2b*, p. 1435 Stuttgart: Thieme 1976
543. Cope, A. C., Holmes, H. L., House, H. O.: Org. React. *9*, 107 (1957)
544. Ueno, Y., Setoi, H., Okawara, M.: Tetrahedron Lett. *1978*, 3753
545. Ksander, G. M., McMurry, J. E.: Tetrahedron Lett. *1976*, 4691
546. Field, G. F., Zally, W. J.: Synthesis *1979*, 295
547. Becher, J.: Synthesis *1980*, 589
548. Troostwijk, C. B., Kellogg, R. M.: Chem. Commun. *1977*, 932
549. Barton, D. H. R. et al.: J. Chem. Soc. (Perkin I) *1977*, 1075
550. Goswami, R.: J. Am. Chem. Soc. *102*, 5973 (1980)
551. Bauer, D. P., Macomber, R. S.: J. Org. Chem. *41*, 2640 (1976)
552. Petragnani, N., Yonashiro, M.: Synthesis *1980*, 710
553. Krapcho, A. P. et al.: J. Org. Chem. *42*, 1189 (1977)
554. Mulzer, J., Brüntrup, G.: Angew. Chem., Int. Ed. Engl. *16*, 255 (1977)
555. Adam, W., Encarnacion, L. A. A.: Synthesis *1979*, 388
556. Fujita, T. et al.: Synthesis *1979*, 310
557. Konen, D. A., Pfeffer, P. E., Silbert, L. S.: Tetrahedron *32*, 2507 (1976)
558. Seebach, D., Pohmakotr, M.: Helv. Chim. Acta *62*, 843 (1979)
559. Bilyard, K. G., Garratt, P. J., Zahler, R.: Synthesis *1980*, 389
560. Hendrick, C. A. et al.: J. Org. Chem. *40*, 8 (1975)
561. Cardillo, G. et al.: J. Org. Chem. *46*, 2439 (1981)
562. Sum, F. W., Weiler, L.: Tetrahedron Lett. *1979*, 707
563. Hubbard, J. S., Harris, T. M.: Tetrahedron Lett. *1978*, 4601
564. Mellor, M., Pattenden, G.: Syn. Commun. *9*, 1 (1979)
565. Pollet, P., Gelin, S.: Synthesis *1978*, 142
566. McMurry, J. E., Andrus, W. A., Musser, J. H.: Syn. Commun. *8*, 53 (1978)
567. Koreeda, M., Liang, Y., Akagi, H.: Chem. Commun. *1979*, 449
568. Taber, D. F.: J. Org. Chem. *41*, 2649 (1976)
569. Birch, A. J., Slobbe, J.: Aust. J. Chem. *30*, 1045 (1977)
570. Caine, D., Frobese, A. S.: Tetrahedron Lett. *1978*, 5167
571. Seebach, D., Lehr, F.: Angew. Chem., Int. Ed. Engl. *15*, 505 (1976); Helv Chim. Acta *62*, 2239 (1979)
572. Shen, C. C., Ainsworth, C.: Tetrahedron Lett. *1979*, 83
573. Wittig, G., Reiff, H.: Angew. Chem., Int. Ed. Engl. *7*, 7 (1968)
574. Cuvigny, T., Larchevêque, M., Normant, H.: Ann. Chem. *1975*, 719
575. Jacobson, R. M., Abbaspour, A., Lahm, G. P.: J. Org. Chem. *43*, 4650 (1978)
576. Larchevêque, M., Valette, G., Cuvigny, T.: Synthesis *1977*, 424
577. Jacobson, R. M., Raths, R. A., McDonald, J. H. III: J. Org. Chem. *42*, 254 (1977)
578. Cuvigny, T. et al.: Synthesis *1978*, 390
579. Larchevêque, M., Valette, G., Cuvigny, T.: Tetrahedron *35*, 1745 (1979)
580. Hashimoto, S., Koga, K.: Tetrahedron Lett. *1978*, 573
581. Worster, P. M., McArthur, C. R., Leznoff, C. C.: Angew. Chem., Int. Ed Engl. *18*, 221 (1979)
582. Enders, D., Eichenauer, H.: Angew. Chem., Int. Ed. Engl. *15*, 549 (1976) Tetrahedron Lett. *1977*, 191; Chem. Ber. *112*, 2933 (1979)
583. Meyers, A. I.: Accts. Chem. Res. *11*, 375 (1978)
584. Sachdev, K.: Tetrahedron Lett. *1976*, 4041
585. Seebach, D., Enders, D.: Chem. Ber. *108*, 1293 (1975)

586. Barton, D. H. R. et al.: J. Chem. Soc. (Perkin I) *1975*, 579
587. Seebach, D., Enders, D., Renger, B.: Chem. Ber. *110*, 1852 (1977)
588. Enders, D. et al.: Org. Syn. *58*, 113 (1978)
589. Seebach, D., Wykypiel, W.: Synthesis *1979*, 423
590. Comins, D. L. et al.: Synthesis *1978*, 309
591. Murata, S., Matsuda, I.: Synthesis *1978*, 221
592. Stork, G., Ozorio, A. A., Leong, A. Y. W.: Tetrahedron Lett. *1978*, 5175
593. Kofron, W. G., Yeh, M. K.: J. Org. Chem. *41*, 439 (1976)
594. Bellassoued, M. et al.: J. Organometal. Chem. *165*, 1 (1979)
595. Bellassoued, M., Dardoize, F., Gaudemar, M.: J. Organometal. Chem. *117*, 35 (1979)
596. Beam, C. F. et al.: J. Org. Chem. *35*, 1806 (1970)
597. Griffiths, J. S., Beam, C. F., Hauser, C. R.: J. Chem. Soc. (C) *1971*, 974
598. Hänssgen, D., Odenhausen, E.: Chem. Ber. *112*, 2389 (1979)
599. Henoch, F. E., Hampton, K. G., Hauser, C. R.: J. Am. Chem. Soc. *91*, 676 (1969)
600. Beam, C. F., Foote, R. S., Hauser, C. R.: J. Chem. Soc. (C) *1971*, 1658
601. Evans, D. A., Sidebottom, P. J.: Chem. Commun. *1978*, 753
602. Tamaru, Y. et al.: J. Am. Chem. Soc. *102*, 7806 (1980)
603. Pohmakotr, M., Geiss, K. H., Seebach, D.: Chem. Ber. *112*, 1420 (1979)
604. Tanaka, K. et al.: Chem. Lett. *1977*, 471
605. Pohmakotr, M., Seebach, D.: Tetrahedron Lett. *1979*, 2271
606. Kirmse, W.: Carbene Chemistry, 2nd Ed., p. 136, New York: Academic Press 1971
607. Normant, H., Castro, B.: C.R. Acad. Sci. *259*, 830 (1964)
608. McLennan, D. J.: Quart. Rev., Chem. Soc. *21*, 490 (1967)
609. Koch, H. F. et al.: J. Am. Chem. Soc. *103*, 5417, 5423 (1981)
610. Casey, C. P., Marten, D. F.: Tetrahedron Lett. *1974*, 925
611. Abdulla, R., Fuhr, K.: J. Org. Chem. *43*, 4248 (1978)
612. Debal, A., Cuvigny, T., Larchevêque, M.: Tetrahedron Lett. *1977*, 3187
613. Barluenga, J., Fañanás, F. J., Yus, M.: J. Org. Chem. *46*, 1281 (1981)
614. Barluenga, J., Yus, M., Bernad, P.: Chem. Commun. *1978*, 847
615. Gurudutt, K. N., Ravindranath, B.: Tetrahedron Lett. *21*, 1173 (1980)
616. Hudrlik, P. F., Kulkarni, A. K.: J. Am. Chem. Soc. *103*, 6251 (1981)
617. Bunnett, J. F.: J. Chem. Educ. *51*, 312 (1974)
618. Reetz, M. T., Schinzer, D.: Angew. Chem., Int. Ed. Engl. *16*, 44 (1977)
619. Brown, H. C. et al.: J. Organometal. Chem. *188*, 1 (1980)
620. Buhler, J. D.: J. Org. Chem. *38*, 904 (1973)
621. Bennett, G. B., Houlihan, W. J., Mason, R. B.: J. Organometal. Chem. *99*, 185 (1979)
622. Gausing, W., Wilke, G.: Angew. Chem., Int. Ed. Engl. *17*, 371 (1978)
623. Clark, T., Schleyer, P. v. R.: Chem. Commun. *1976*, 798
624. Pines, H., Eschinazi, H. E.: J. Am. Chem. Soc. *78*, 5950 (1956)
625. Skinner, D. L., Peterson, D. J., Logan, T. J.: J. Org. Chem. *32*, 105 (1967)
626. Hart, H., Curtis, O. E.: J. Am. Chem. Soc. *79*, 931 (1957)
627. Bordwell, F. G.: Accts. Chem. Res. *3*, 281 (1970)
628. Philips, J. C. et al.: Chem. Commun. *1971*, 22
629. Shapiro, R. H.: Org. React. *23*, 405 (1976)
630. Kolonko, K. J., Shapiro, R. H.: J. Org. Chem. *43*, 1404 (1978)
631. Chamberlain, A. R., Bond, F. T.: Synthesis *1979*, 44
632. Taylor, R. T. et al.: Tetrahedron Lett. *1977*, 159

633. Stemke, J. E., Chamberlain, A. R., Bond, F. T.: Tetrahedron Lett. *1976*, 2947
634. Traas, P. C., Boelens, H., Takken, H. J.: Tetrahedron Lett. *1976*, 2287
635. Nakai, T., Mimura, T.: Tetrahedron Lett. *1979*, 531
636. Kano, S. et al.: Synthesis *1978*, 305
637. Bartmess, J. E. et al.: J. Am. Chem. Soc. *103*, 4746 (1981)
638. Neukam, W., Grimme, W.: Tetrahedron Lett. *1978*, 2201
639. Wittig, G., Tochtermann, W.: Chem. Ber. *94*, 1692 (1961)
640. Schöllkopf, U., Hoppe, I.: Angew. Chem., Int. Ed. Engl. *14*, 765 (1975)
641. Bank, S. et al.: J. Org. Chem. *46*, 1243 (1981)
642. Guthrie, R. D. in Buncel, E., Durst, T.: Comprehensive Carbanion Chemistry, p. 197, New York: Elsevier 1980
643. Guthrie, R. D. et al.: J. Org. Chem. *46*, 498 (1981)
644. Bailey, E. J. et al.: J. Chem. Soc. *1962*, 1578
645. Adam, W., Cueto, O.: J. Org. Chem. *42*, 38 (1977)
646. Wasserman, H. H., Lipshutz, B. H.: Tetrahedron Lett. *1975*, 1731
647. Gardner, J. N., Carlon, F. E., Gnoj, O.: J. Org. Chem. *33*, 3294 (1968)
648. Vedejs, E., Engler, D. A., Telschow, J. E.: J. Org. Chem. *43*, 188 (1978)
649. Selikson, S. J., Watt, D. S.: J. Org. Chem. *40*, 267 (1975)
650. Williams, J. R., Unger, L. R., Moore, R. H.: J. Org. Chem. *43*, 1271 (1978)
651. Wasserman, H. H., Lipshutz, B. H.: Tetrahedron Lett. *1976*, 4613
652. Eglinton, G., McCrae, W.: Adv. Org. Chem. *4*, 225 (1963)
653. Kornblum, N., Cheng, L.: J. Org. Chem. *42*, 2944 (1977)
654. Kobayashi, Y., Taguchi, T., Tokuno, E.: Tetrahedron Lett. *1977*, 3741
655. Ito, Y. et al.: J. Am. Chem. Soc. *99*, 1487 (1977)
656. Padwa, A. et al.: J. Org. Chem. *44*, 3281 (1979)
657. Galakatos, N. et al.: Synthesis *1978*, 472
658. Yasuda, H., Tani, H.: Tetrahedron Lett. *1975*, 11
659. Hsu, H. F.: Ph. D. Thesis, University of Arizona, 1982
660. Ogle, C. A.: Ph. D. Thesis, University of Arizona, 1982
661. Ogle, C. A.: M. S. Thesis, University of Arizona, 1980
662. Gausing, W., Wilke, G.: Org. Syn., in press
663. Kershner, L. D., Gaidis, J. M., Freedman, H. H.: J. Am. Chem. Soc. *94*, 985 (1972)
664. Hauptmann, H.: Angew. Chem., Int. Ed. Engl. *14*, 498 (1975)
665. Fu, P. F., Harvey, R. G.: Chem. Rev. *78*, 317 (1978)
666. Laguerre, M. et al.: J. Org. Chem. *44*, 4275 (1979)
667. Grossert, J. S. et al.: Tetrahedron Lett. *1974*, 2805
668. Kauffmann, T. et al.: Chem. Ber. *111*, 1330 (1978)
669. Boche, G., Bieberbach, A.: Tetrahedron Lett. *1976*, 1021
670. Mackenzie, D. in Zabicky, J.: Chemistry of Alkenes 2, p. 115, New York: Interscience 1970
671. Brown, C. A.: Synthesis *1978*, 754
672. Mills, N. S., Shapiro, J., Hollingsworth, M.: J. Am. Chem. Soc. *103*, 1263 (1981)
673. Wotiz, J. H. in Viehe, H. G.: Chemistry of Acetylenes, p. 371, New York: Marcel Dekker, 1969
674. Hommes, H., Brandsma, L.: Recl. Trav. Chim. *96*, 160 (1977)
675. Kossa, W. C., Rees, T. C., Richey, H. G.: Tetrahedron Lett. *1971*, 3455
676. Felkin, H. et al.: Tetrahedron Lett. *1972*, 2285
677. Bates, R. B. et al.: J. Org. Chem. *45*, 168 (1980)
678. Jung, M. E.: Tetrahedron *32*, 1 (1976)

679. Corbel, B., Durst, T.: J. Org. Chem. *41*, 3648 (1976)
680. Stirling, C. J. S.: Chem. Rev. *78*, 517 (1978)
681. Hill, E. A., Davidson, J. A.: J. Am. Chem. Soc. *86*, 4663 (1964)
682. Robinson, R., Smith, L. H.: J. Chem. Soc. *1936*, 195
683. Paul, R., Tchelitcheff, S.: Bull. Soc. Chim. France *1952*, 808
684. Jacobi, P. A., Ueng, S., Carr, D.: J. Org. Chem. *44*, 2042 (1979)
685. Vinick, F. J., Pan, Y., Gschwend, H. W.: Tetrahedron Lett. *1978*, 4221
686. Taschner, M. J., Kraus, G. A.: J. Org. Chem. *43*, 4235 (1978)
687. Gronowitz, S., Frejd, T.: Acta Chem. Scand. (B) *29*, 818 (1975)
688. Astsumi, K., Kuwajima, I.: J. Am. Chem. Soc. *101*, 2208 (1979)
689. Grovenstein, E.: Angew. Chem., Int. Ed. Engl. *17*, 313 (1978)
690. Pine, S. H.: Org. React. *18*, 403 (1970)
691. Jenny, E. F., Druey, J.: Angew. Chem., Int. Ed. Engl. *74*, 155 (1962)
692. Dolling, U. H. et al.: Chem. Commun. *1975*, 545
693. Baldwin, J. E., Patrick, J. E.: J. Am. Chem. Soc. *93*, 3556 (1971)
694. Felkin, H., Frajerman, C.: Tetrahedron Lett. *1977*, 3485
695. Chérest, M., Felkin, H., Frajerman, C.: Tetrahedron Lett. *1977*, 3489
696. Bates, R. B. et al.: J. Am. Chem. Soc. *92*, 6345 (1970)
697. Klein, J., Glily, S.: Tetrahedron *27*, 3477 (1971)
698. Durst, T., van den Elzen, R., Le Belle, M.: J. Am. Chem. Soc. *94*, 9261 (1972)
699. Vedejs, E. et al.: J. Org. Chem. *43*, 4831 (1978)
700. Cazes, B., Julia, S.: Bull. Soc. Chim. France *1977*, 925
701. Cazes, B., Julia, S.: Tetrahedron *35*, 2655 (1979)
702. Wada, M. et al.: Chem. Lett. *1977*, 557
703. Ceré, V. et al.: J. Org. Chem. *43*, 4826 (1978)
704. Schoufs, M. et al.: Synthesis *1978*, 439
705. Kloosterziel, H., van Drunen, J. A. A.: Recl. Trav. Chim. *88*, 1084 (1969)
706. Paquette, L. A., Crouse, G. D.: J. Am. Chem. Soc. *103*, 6235 (1981)
707. Huisgen, R.: Angew. Chem., Int. Ed. Engl. *19*, 947 (1980)
708. Mulvaney, J. E., Savage, D.: J. Org. Chem. *36*, 2592 (1971)
709. Staley, S. W., Dustman, C. K., Linkowski, G. E.: J. Am. Chem. Soc. *103*, 1069 (1981)
710. Bates, R. B., McCombs, D. A.: Tetrahedron Lett. *1969*, 977
711. Hunter, D. H., Sim, S. K.: J. Am. Chem. Soc. *91*, 6202 (1969)
712. Ficini, J., Claeys, M., Depezay, J. C.: Tetrahedron Lett. *1973*, 3353
713. Kloosterziel, H., van Drunen, J. A. A., Galama, P.: Chem. Commun. *1969*, 885
714. Kloosterziel, H., van Drunen, J. A. A.: Recl. Trav. Chim. *89*, 667 (1970)
715. Magid, R. M., Wilson, S. E.: Tetrahedron Lett. *1971*, 19
716. Grovenstein, E., Jr., Chiu, K. W., Patil, B. B.: J. Am. Chem. Soc. *102*, 5848 (1980)
717. Wilson, S. E.: Tetrahedron Lett. *1975*, 4651
718. Freeman, P. K. et al.: J. Org. Chem. *32*, 3958 (1967)
719. Freeman, P. K., George, D. E., Rao, V. N. M.: J. Org. Chem. *28*, 3234 (1963)
720. Staley, S. W., Erdman, J. P.: J. Am. Chem. Soc. *92*, 3832 (1970)
721. Bates, R. B., Brenner, S., Cole, C. M.: J. Am. Chem. Soc. *94*, 2130 (1974)
722. Schell, F. M., Carter, J. P., Wiaux-Zamar, C.: J. Am. Chem. Soc. *100*, 2894 (1979)
723. Kuwajima, I., Uchida, M.: Tetrahedron Lett. *1972*, 649
724. Sowerby, R. L., Coates, R. M.: J. Am. Chem. Soc. *94*, 4758 (1972)
725. Corey, E. J., Jautelat, M.: Tetrahedron Lett. *1968*, 5787

726. Kauffmann, T., Joussen, R., Woltermann, A.: Angew. Chem., Int. Ed. Engl. *16*, 709 (1977)
727. Martin, S. F.: Synthesis *1979*, 633
728. Hase, T. A., Koskimies, J. K.: Aldrichimica Acta *14*, 73 (1981); *15*, 35 (1982)
729. Gokel, G. W. et al.: Tetrahedron Lett. *1979*, 3375
730. Soderquist, J. A., Hsu, G. J. H.: Organometallics *1*, 830 (1982)
731. Abramovitch, R. A. et al.: Tetrahedron Lett. *21*, 705 (1980)
732. Kocienski, P. J.: Tetrahedron Lett. *21*, 1559 (1980)
733. Merz, A.: Synthesis *1974*, 724
734. Heissler, D. et al.: Tetrahedron Lett. *1976*, 4879
735. Jung, M. E., Blum, R. B.: Tetrahedron Lett. *1977*, 3791
736. Stork, G., Leong, A. Y. W., Touzin, A. M.: J. Org. Chem. *41*, 3491 (1976)
737. Kalir, A., Balderman, D.: Synthesis *1973*, 358
738. Deuchert, K., Hertenstein, U., Hünig, S.: Synthesis *1973*, 777
739. Hünig, S., Wehner, G.: Synthesis *1975*, 180
740. Baldwin, J. E., Höfle, G. A., Lever, O. W.: J. Am. Chem. Soc. *96*, 7125 (1974)
741. Niznik, G. E., Morrison, W. H., Walborsky, H. M.: J. Org. Chem. *39*, 600 (1974)
742. Seebach, D., Corey, E. J.: J. Org. Chem. *40*, 231 (1975)
743. Hünig, S., Wehner, G.: Synthesis *1975*, 391
744. Stork, G., Depezay, J. C., d'Angelo, J.: Tetrahedron Lett. *1975*, 389
745. Ogura, K. et al.: Tetrahedron Lett. *1974*, 3653
746. Blatcher, P., Grayson, J. I., Warren, S.: Chem. Commun. *1976*, 547
747. van Leusen, D., van Leusen, A. M.: Tetrahedron Lett. *1977*, 4233
748. Schmidt, R. R., Schmid, B.: Tetrahedron Lett. *1977*, 3583
749. Ager, D. J.: Tetrahedron Lett. *21*, 4759 (1980)
750. Anderson, R. J., Henrick, C. A.: J. Am. Chem. Soc. *97*, 4327 (1975)
751. Still, W. C., Macdonald, T. L.: J. Am. Chem. Soc. *96*, 5561 (1974)
752. Kondo, K., Saito, E., Tunemoto, D.: Tetrahedron Lett. *1975*, 2275
753. Giess, K. H., Seebach, D., Seuring, B.: Chem. Ber. *110*, 1833 (1977)
754. Caine, D., Frobese, A.: Tetrahedron Lett. *1978*, 883
755. Seyferth, D., Mammarella, R. E., Klein, H. A.: J. Organometal. Chem. *194*, 1 (1980)
756. Julia, M., Ward, P.: Bull. Soc. Chim. France *1973*, 3065
757. Kondo, K., Matsumoto, M.: Tetrahedron Lett. *1976*, 391
758. Duhamel, L., Poirier, J. M.: J. Org. Chem. *44*, 3585 (1979)
759. Evans, D. A., Andrews, G. C., Buckwalter, B.: J. Am. Chem. Soc. *96*, 5560 (1974)
760. Still, W. C., Macdonald, T. L.: J. Org. Chem. *41*, 3620 (1976)
761. Baba, S., Van Horn, D. E., Negishi, E.: Tetrahedron Lett. *1976*, 1927
762. Raucher, S., Koolpe, G. A.: J. Org. Chem. *43*, 4252 (1978)
763. Evans, D. A. et al.: Tetrahedron Lett. *1973*, 1385, 1389
764. Corey, E. J., Erickson, B. W., Noyori, R.: J. Am. Chem. Soc. *93*, 1724 (1971)
765. Wada, M. et al.: Chem. Lett. *1977*, 345
766. Yu, L. C., Helquist, P.: Tetrahedron Lett. *1978*, 3423
767. Kimura, R., Ichihara, A., Sakamura, S.: Synthesis *1979*, 516
768. Jenkitkasemwong, Y., Thebtaranonth, Y., Wajirum, N.: Tetrahedron Lett. *1979*, 1615
769. Ichihara, A., Nio, N., Sakamura, S.: Tetrahedron Lett. *1980*, 4467
770. Rollinson, S. W., Amos, R. A., Katzenellenbogen, J. A.: J. Am. Chem. Soc. *103*, 4114 (1981)
771. Magnus, P. D.: Tetrahedron *33*, 2019 (1977)

772. Clinet, J. C., Linstrumelle, G.: Tetrahedron Lett. *1978*, 1137
773. Cazes, B. et al.: J. Organometal. Chem. *177*, 67 (1979)
774. Itoh, A. et al.: Tetrahedron Lett. *21*, 361 (1980)
775. Boeckman, R. K., Bruza, K. J.: Tetrahedron Lett. *1977*, 4187
776. Reich, H. J., Gold, P. M., Chow, F.: Tetrahedron Lett. *1979*, 4433
777. Oshima, K., Yamamoto, H., Nozaki, H.: J. Am. Chem. Soc. *95*, 4446 (1973)
778. Oshima, K. et al.: J. Am. Chem. Soc. *95*, 2693 (1973)
779. Gill, M., Bainton, H. P., Rickards, R. W.: Tetrahedron Lett. *22*, 1437 (1981)
780. Cavill, G. W. K., Goodrich, B. S., Laing, D. G.: Aust. J. Chem. *23*, 83 (1970)
781. Stetter, H.: Angew. Chem. *67*, 769 (1955)
782. Mori, K., Uchida, M., Matsui, M.: Tetrahedron *33*, 385 (1977)
783. Piers, E., Grierson, J. R.: J. Org. Chem. *42*, 3755 (1977)
784. Cohen, T., Bennett, D. A., Mura, A. J.: J. Org. Chem. *41*, 2506 (1976)
785. Amupitan, J., Sutherland, J. K.: Chem. Commun. *1978*, 852

Subject Index

Acenaphthene dianion dilithium 8
Acetaldehyde acidity of 21
Acetamide acidity of 17
Acetic acid acidity of 21
Acetic anhydride carbanion acetylation with 25
Acetoacetic ester syntheses 22, 48, 49
Acetone acidity of 17, 21
  addition of carbanions to 40
  cyanoethylation of 35
Acetonitriles acidity of 17, 20, 21
  from metallated acetonitriles 54
  metallated reactions with electrophiles 54
  metallation of 23, 54, 61, 62
Acetophenone acidity of 17
Acetylacetonate lithium 11
Acetylacetone acidity of 17
Acetylenes see Alkynes
Acetylenic anions from isomerizations of alkynes and allenes 68
  from terminal alkynes 19
  reactions of 38, 39, 64
  structure of 7
Acid chlorides reactions with Cd, Cu, Mn, and Rh reagents 32, 33
  reactions with dianions 50, 51
  reactions with enolates 45
Acidities comparison between solution and gas phase, 20, 21
  gas phase 20, 21
  in DMSO 17–20
  kinetic 20
  of CH protons 17–21
  range of 21
  thermodynamic 17–20
Acrylic acids from Wittig-type reactions 37
Acrylonitrile anionic polymerization of 34
  nucleophilic additions to 35

Acyl carbanions equivalents of 75
  from carbanions and CO 30, 38
  instability of 38
Acyl cyanides reaction with enolates 45
Acyl groups carbanion-stabilizing ability of, 18
Additions of carbanions 34–39
  acetylenic 38, 39
  aryl 38
  to aldehydes 30, 36, 37, 50–52, 56, 58, 59
  to alkenes 26, 30, 34, 35, 37, 48, 51, 52, 58, 68, 73, 74
  to alkynes 30, 32, 35, 48, 73
  to arenes 59
  to benzynes 47
  to carbon dioxide 30, 38
  to carbon disulfide 30, 38, 50, 51
  to carbon monoxide 30, 38
  to carboxylates 38
  to cycloheptatriene 26, 34
  to cyclooctatetraene 34
  to 1,3-dienes 34
  to esters 30, 38
  to imines 30, 35, 36, 40, 47
  to iminium salts 30, 36, 47
  to ketones 30, 36, 37, 39, 40, 56
  to nitriles 30, 36
  to pyridinium salts 47
  vinyl 38
1,2-Additions vs. 1,4-additions to α,β-unsaturated carbonyl compounds 34, 35, 38
Aggregation of organometallics 3, 20, 43
Alcohols as solvents for carbanions 22
  from carbanions 29, 30, 32, 36–38, 46, 47, 56, 63
  solution and gas phase acidities of 20

Aldehydes additions of carbanions to 28, 30, 36, 37, 45–47, 49, 51, 52, 56, 58, 59
  conversion to tosylhydrazones 61
  from carbanions and DMF 32, 33
  from carbanions and formates 32, 33
  from carbanions *via* dithioacetals 30, 31, 38
  from hydrolysis of α-aminonitriles 54
  from hydrolysis of imines 52
  from hydrolysis of vinyl sulfides 61
  from α-silylepoxides 37
  α-halo-, from α-halosulfoxides 37
  β-hydroxy-, dehydration of 45–47
    from aldol condensations 45–47
  in transaminations 42
  reaction with amines 47
  α,β-unsaturated from aldol condensations 45–47
  from vinyl anions 61
Aldol condensations base-catalyzed 22, 44–47, 53, 54, 58, 68
  reverse 69
  stereochemistry of 46, 47
Alkali metals *see* also individual metals
  metallation of acidic hydrocarbons with 23
  reduction of 1,3-dienes with 25
Alkanes acidity of 19, 21
  as solvents for carbanions 22
  preparation of symmetrical 64
  preparation of unsymmetrical 29, 30
Alkenes acidity of 19
  by elimination of alkoxides 69
  by elimination of carboxylates 49, 59
  by elimination of oxide ion 58
  by elimination of SO$_2$ 66
  by elimination of trimethylsiloxide 59
  by Wittig reactions 37
  conversion to carbanions through mercuriation 27
  from allyl alcohols 31
  from allyl anions and electrophiles 40–42, 67
  from carbanions by hydride transfer 36
  from cyclopentyl anions 61

from dithiocarbamates 31
from enolates 45
from Horner-Emmons reactions 37
from Peterson reactions 37
from pyrolysis of amines oxides 40
from pyrolysis of α-keto-γ-lactones 49
from pyrolysis of β-lactones 50
from pyrolysis of selenoxides and sulfoxides 31
from vinyl carbanions and electrophiles 35, 38
metallation of 19, 24, 41, 60, 67
positional isomerization of 41, 42, 67
Alkoxides *see* also Potassium *t*-butoxide
  as bases for carbanion preparation 22, 36, 37, 58, 68, 69
  as leaving groups in eliminations from carbanions 57, 58
  as protecting groups for hydroxyls 58
  from Witting rearrangements 70
  reaction with trimethylsilyl chloride 45
α-Alkyl groups stabilizing and destabilizing effects on carbanions of 18, 21
Alkyl halides metallations of 18, 19
  reactions with: carbanions 29–31, 50–56, 65, 66
    copper-lithium reagents 29
    iron carbonyl complexes 33
  reactivity as a function of halogen 14, 15
Alkyl isothiocyanates reactions with carbanions 50
Alkyl phenyl sulfides reactions with lithium 15
Alkylboranes from carbanions 29
  rearrangements of 29
Alkyllithiums *see* also Organolithiums and individual alkyllithiums
  activation by alkoxides 24
  activation by THF 24
  activation by TMEDA 24
  as bases for carbanion preparation 21, 23
  cleavage of solvents by 24
  determination of concentration of 25

reactions with: aldehydes 36
 alkenes 34
 alkyl halides 29
 amides 32
 carboxylic acids 32
 cuprous halides 29
 D_2O 29
 epoxides 31
 vinyl halides 32
 relative metallation rates with, 20
Alkylpotassiums non-activation by TMEDA 24
Alkylsodiums activation by TMEDA 24
 reactions with alkyl halides 29
Alkylthio groups, carbanion-stabilizing ability of 18
Alkynes additions to 30, 35, 48
 as intermediates between vinyl halides and alkenes 32, 48
 from alkenes 60
 from allenyl halides 30, 31
 from dianions 50, 52
 from other alkynes 23, 38, 39, 64, 68, 73
 from sulfones 60
 from vinyl anions 57, 60, 61, 69
 from vinyl halides 60
 isomerization of 23, 68
 structures of 10
 terminal metallation of 19, 21, 23
Allenes acidity of 10
 from dianions 52
Allenyl ethers from propargyl acetals 30, 31
Allenyl halides conversion to alkynes 30, 31
Allyl alcohols γ-alkylation of 30, 31
Allyl anions additions to aldehydes and ketones of 28, 40, 42, 46, 47
 additions to alkenes of 74
 additions to imines of 40
 eliminations of alkyl anions from 73, 74
 eliminations of hydride ions from 73
 equilibration of 67
 failure to undergo sigmatropic [1,4]-proton shifts of 70
 geometry of 1-alkyl substituents in 9
 oxidative couplings of 64

 position of attack by electrophiles in unsymmetrical 40–42
 preparations of 21, 22, 24, 25, 27, 34, 41, 42, 57, 67, 72–74
 rotations in 9, 10, 63, 67
 structures of 8–10, 14, 15
 thermal instability of 70
 3-vinyl- sigmatropic rearrangement of 74
 with quaternary ammonium counterions 28
Allyl halides in Reformatsky reactions 15
 reactions with carbanions 30, 31, 52, 53
Allyl phenyl ethers reactions with lithium 15
Allyltrimethylsilane as a precursor for allyl anion 28
Amide ions as leaving groups in eliminations from carbanions 57, 58
 as protecting groups for amines 58
 as strong bases 23
 from carbanions and imines 47
 reactions with electrophiles 55
Amides additions of carbanions to α,β-unsaturated 35
 metallation of 18, 19
 reactions with carbanions of 32, 33, 38
Amidines as bases 23
 from carbanions 73
Amines by additions of carbanions to imines 30, 35, 36, 40
 by additions of carbanions to iminium salts 35, 36
 from N-nitrosamines 54
 in Mannich reactions 47
 in promoting reactions of Grignard reagents with alkyl halides 31
 metallation of 18–20
 primary from reductions of nitro compounds 35
 in transaminations 42
 tertiary from carbanions 30, 33
 from sigmatropic rearrangements 71
 from Stevens rearrangements 70
 vinyl, from Wittig-type reactions 37
 metallations of 19
Ammonia acidity of 17, 21
 as a solvent for carbanions 23

Amylsodium   15
Anhydrides reactions with carbanions 32, 45, 56
Anion radicals as intermediates in $S_{RN}1$ reactions   44
   cleavages to anions and radicals   26, 29, 44
   dimerizations of   65, 66
   from oxidations of dicarbanions   65, 66
   from reductions of alydehydes and ketones   36
   from reductions of sigma bonds   26
   from reductions of unsaturated compounds   25, 36
Anthracene dianion preparations of 8, 25
   structure of   8
Antiaromaticity of benzene dianion   25
   of cyclooctatetraene   25
Aromaticity in stabilizing carbanions 18, 25
Aromatics metallation of   19, 20
   ortho acylation of   71
Aryl carbanions preparations of   13, 16, 24, 27, 47
   reactions of   36, 38, 47
   structures of   7
Aryl groups carbanion-stabilizing ability of   18
Aryl halides in $S_{RN}1$ reactions   44
   reactions with: enolates   44
      metals   14
      nucleophiles   59
      organometallics   32, 38
      transition metal compounds   44
Asymmetric syntheses aldol type   46, 47
   of α-alkylated acids   54
   of α-alkylated aldehydes and ketones 53, 54
Azines from dimetallated azines   55
Aziridines preparation of   47
Azobisisobutyronitrile   28

Barbier reactions   14, 36
Barium reaction with alkyl halides   15
Bases hindered tertiary amine   23
Base-solvent systems   21–25
Benzene acidity of   21
   from hydride elimination   59

Benzene dianion, antiaromaticity of 25
Benzoic acid acidity of   17, 21
Benzonitrile from ring opening of metallated isoxazoles   61
Benzophenone dianion dilithium   11
2-O-Benzothiazole as a leaving group 31
   on an allyl anion   40
Benzyl anions preparations of   15, 21, 22, 24, 26, 27, 66, 71–73
   rearrangements of   71–73
   structures of   8, 14
Benzyne intermediates   32, 47
Bicyclo[1.1.0]butan-1-yllithium   3, 6
Biphenyls from carbanions   73
Bis(diphenylphosphino)methane acidity of   17
Bis(phenylthio)methane acidity of   17
Bond lengths   5
Bond orders   10, 11
Bond strengths carbon-metal   3, 4, 6–8, 11
Bonds 2-center   3–8
   3-center-2-electron   3–5, 7
   4-center-2-electron   3–7
   metal-metal   5, 6
   pi   4, 7–12
   sigma   3–8
Boranes alkyl   29, 38, 41, 59, 63
   dialkoxyfluoro   29
   trialkoxy   29
Borate trimethyl reaction with carbanions   41
Bordwell's rule   31, 63
Bridging alkyl groups   3, 5, 7
Bridging halogens   5
N-Bromosuccinimide reactions with carbanions   30, 33, 38
Butadiene anionic polymerization and telomerization of   34
   from oxidation of butadiene dianion 65
   reduction to butadiene dianion   25
Butadiene dianions from electrocyclizations   73
   hetero analogs of   55, 56
   hydride eliminations from   73
   isomerization of   68
   oxidation to butadienes of   65
   preparation by dimetallations   68
   preparation from butadienes   25

structure of 9
2-vinyl- oxidative coupling of 65
t-Butanol acidity of 17, 20, 21
  as a solvent for carbanion formation
  22
2-Butenes, metallation of 41
1-Buten-2-olate lithium 3,3-dimethyl-
  10
t-Butoxides as bases for carbanion preparations 21, 22, 24
t-Butyl bromide reactions with carbanions 31
t-Butyl ether 14
n-Butyllithium activation by potassium
  t-butoxide 24, 42
  activation by THF 24
  activation by TMEDA 24
  as as base for carbanion preparations
  16, 18, 20, 23, 24
  in transmetallations 28
  structure of 5, 6
sec-Butyllithium as a superior metallating agent 20, 27
t-Butyllithium 5, 16
n-Butylpotassium eliminations from
  60

Cadmium chloride oxidation of carbanions with 66
Calcium reaction with alkyl halides 15
Carbanion equivalents 75, 76
Carbanions conversions to: alcohols
  29–32, 36–42, 47, 54, 63, 64, 68–70
  aldehydes 32, 33, 52–55, 61
  alkanes 29, 30, 64
  alkenes 30, 31, 35, 36, 45, 50,
    57–61, 66
  alkylboranes 29, 30, 63
  alkynes 30, 35, 50, 52, 57, 60,
    61, 64, 69, 73
  allenes 30, 31, 52
  amidines 73
  amines 30, 33, 35, 47, 54–56, 70,
    71
  α-aminoketones 56
  β-aminoketones 47
  arenes 59, 60, 73
  azines 55
  aziridines 47
  carbenes 57, 58
  carbon radicals 63
  carbonium ions 63

carboxylic acids 30, 38, 49–51,
  54
cyclobutanes 69
cycloheptanes 74
cyclooctatetraenes 34, 66
cyclopentanes 73, 74
cyclophanes 66
cyclopropanes 48, 57, 60, 66, 69
1,3-dienes 65
1,5-dienes 64
1,2-diketones 65
1,3-diketones 45, 51
1,4-diketones 53, 64
1,5-diketones 53
dithioacetals 30, 38
epoxides 47, 66
esters 33, 44, 51, 52
furans 73
halides 30, 33
hydrazones 55
hydroperoxides 63, 64
β-hydroxyacids 50
β-hydroxyaldehydes 45–47
β-hydroxyketones 45–48
imines 52, 53, 64
indoles 44
isocyanides 69
isoxazoles 55
isoxazolines 55
β-ketoacids 50
β-ketoesters 45, 49, 51
ketones 29, 30, 32, 33, 36, 38,
  39, 47, 49, 53, 64
β-lactones 50
γ-lactones 52
mercaptans 31, 33
nitriles 30, 33, 54, 61, 69
nitro compounds 30, 33, 35, 52,
  64
nitroates 52
α-nitroketones 49
N-nitroso compounds 54
oxazines 54
oxazolines 54
oximes 55
phosphines 30, 33
phosphonates 54
pyrazoles 55
pyridines 47
quinones 65
silanes 30, 33
sulfides 30, 33, 56, 61, 70, 71

sulfones 30, 33
thiazoles 54
thioamides 56
thioesters 51, 56
α-thiopyridones 49
α,β-unsaturated acids 51
α,β-unsaturated aldehydes 45–47, 58, 61
α,β-unsaturated esters 45, 46, 48, 49, 56
β,γ-unsaturated esters 48
α,β-unsaturated ketones 45–48, 52–54, 60
β,γ-unsaturated ketones 48, 51, 71
vinyl siloxanes 48
vinyl sulfides 56, 70
xylyenes 66
ease of oxidation of 31, 63
from alkanes by proton abstraction 17–25
from alkyl halides and metals 13–15
from alkyl halides and organometallics 15, 16
from organometallics and metal salts 27, 28
from organometallics and metals 27
from organometallics by rearrangement 67–74
from unsaturated compounds and metals 25, 26
from unsaturated compounds and organometallics 26
halide-free 27
importance of 1, 77
preparations of 13–28, 67–74
reactions with: acid chlorides 45, 50, 51
acids 29, 30
acyl cyanides 45
acyl halides 45
acylating agents 30, 32, 45
alkyl halides 29, 30, 43, 50–56, 58, 60, 66
alkyl nitrates 30, 33
allyl halides 30, 31, 52, 53
amides 32, 33, 38, 61
anhydrides 45, 56
aryl halides 32, 38, 44
boranes 29, 30, 38, 41
cadmium ion 66

carbenes 40
carbon dioxide 29, 38, 55
carbon disulfide 29, 38
carbon monoxide 30, 38
carbon tetrahalides 30, 33
carbonates 33
carboxylic acids 32, 38
ceric ion 64
chloroboranes 63
chlorophosphines 30, 33
cupric ion 64, 66
cuprous halides 29, 64
cuprous thiophenoxide 34
N,N-dialkyl-O-arenesulfonylhydroxylamines 30, 33
1,2-dibromoethane 65, 66
1,2-dichloroethane 65
disulfides 30, 33, 61
electrophiles 29–56
epoxides 30, 31, 51, 54
esters 30, 32, 38, 39, 45, 49,
ferric ion 64, 66
ferrous ion 40
halogens 30, 33, 64, 66
N-halosuccinimides 30, 33, 38 39
hexahaloethanes 30, 33
imines 30, 35, 36
iminium salts 30, 36
iron pentacarbonyl 33
isothiocyanates 50
ketones 59
methylene halides 30, 33
molybdenum peroxide 63
nitrobenzene 66
oxygen 30, 33, 63, 64, 66
phenyl arenesulfonates 30, 33
phosphoryl halides 45, 54
propargyl halides 30, 31
pyridines 36
silyl halides 30, 33, 43, 54
singlet oxygen 64
sulfenyl chlorides 30, 33
sulfonyl chlorides 30, 33
sulfur 30, 33
thiocyanates 30, 33
tosyl cyanide 30, 33
vinyl halides 30, 32
zinc ion 64
structures of 3–12
Carbanion-stabilizing abilities additi
effects of 18, 19

increase with increasing character 19

of various groups   17–21, 60

Carbenes 1,4-additions to 1,3-dienes of 40

chloro- reactions with cyclopentadienyl anion   40

from 1,1-eliminations of carbanions 40, 57, 58

Carbon dioxide from α-hydroperoxy-acids   64

reactions with carbanions   30, 38, 55

Carbon disulfide reactions with carbanions   30, 38, 50, 51

Carbon monoxide reactions with carbanions   30, 38

Carbon tetrahalides reactions with carbanions   30, 33

Carbonates dialkyl reactions with carbanions   33

Carbonium ions from carbanion oxidations   63

Carbonyl compounds α,β-unsaturated see also Aldehydes α,β-unsaturated etc. from aldol condensations   22, 44–47, 53, 54, 58

nucleophilic additions to   34, 35, 38

Carbonyl transposition   61

Carboxylates as leaving groups   45, 46, 49, 58, 59

Carboxylic acids dimetallation of   50

from acetic acid dianion   50, 51

from carbanions and CO_2 30, 38

from enaminate-type anions   54

from malonic ester syntheses   49

from 1,1,1-trihalides   31

α-hydroperoxy- from carboxylic acid dianions   64

β-hydroxy- conversion to β-lactones 50

from acetic acid dianion   50

β-keto- from acetic acid dianion   50

metallations of   20

reactions with carbanions   30, 38

α,β-unsaturated from dianions   51

β,γ-unsaturated from dianions   51

Cephalotaxine from an enolate   44

Ceric ion oxidation of carbanions by 64

Charge delocalization   8

Chelation in carbanion stabilization 18, 19, 43, 55

Chichibabin reactions   26

N-Chlorosuccinimide reaction with carbanions   30, 33, 39

CIDNP   14, 16

Claisen condensations   45

$(CLi_4)_n$, 10

$(C_2Li_2)_n$, 10

$(C_2Li_4)_n$, 10

$(C_3Li_4)_n$, 10

Concentration of carbanions measurement of   25

Configurational stability   7, 41, 67, 72, 73

Copper catalyst   32

Copper complexes   31

Copper-lithium reagents see lithium dialkylcuprates

Counterion effects see also lithium dialkylcuprates

in additions of enolates to aldehydes and ketones   46

in alkylations of dianions   52, 56

in alkylations of dienolates   48

in dissociations of ion pairs   67

in eliminations of oxide ion from carbanions   58

in reactions of enolates at C vs. O   43

in ring openings of epoxides   69

in stabilizing carbanions   20–22, 41

Covalent radii of carbon and metals   4

Crotyl anions equilibrium composition of   41

position of attack by electrophiles in 24

preparation of   24, 41

preservation of configuration in   24

Crown ethers   8, 43

Cryptands   11, 43

Cuprates see lithium dialkylcuprates

Cupric ion, oxidative coupling of anions, with   64, 66

Cuprous ion effect on alkylation of dianions   52

in promoting 1,4-additions of carbanions to α,β-unsaturated carbonyl compounds   34, 38

in promoting α-alkylation of prenyl Grignard reagents   41

in promoting γ-alkylation of dienolates   41

in promoting Grignard reactions with alkyl halides and allenyl ethers 31
in promoting nucleophilic additions to terminal alkynes 35
oxidative coupling of enaminates with 64
reaction with alkyllithiums 29, 34
Cyanide ion 7
Cyano group carbanion-stabilizing ability of 18, 20, 71
Cyanoacetic ester syntheses 22
Cyanoethylation 35
Cyanohydrins conversion to ketones 64, 71
from α-metallated nitriles 64, 71
Cyclobutanols synthesis of 47, 69
1,3-Cycloheptadiene metallation of 42
preparation of 42
1,4-Cycloheptadiene metallation of 42
preparation of 42
Cycloheptadienyl anions preparation of 42, 72, 74
protonation of 42
[1,6]sigmatropic proton shifts in 70
Cycloheptatriene addition of carbanions to 26, 34
Cycloheptatrienyl trianions preparation of 34
Cycloheptene 1-methyl- metallation of 60
Cyclohexadienyl anions electrocyclic ring closure of 74
eliminations from 34, 59, 60
from additions of nucleophiles to aromatics 34, 59
from additions of nucleophiles to 3-methylene-1,4-cyclohexadienes 34
structures of 9
Cyclohexane 1,1,4,4-tetravinyl- from dianion alkylation 65
Cyclohexanones acidity of 17
alkylation of enolates of 44
preparation of 43, 68
2-Cyclohexenones from aldol condensations 53, 54
2-metallated equivalents of 48, 52
Cyclohexyllithium structure of 3, 6
1,5-Cyclooctadiene metallation of 60
Cyclooctadienyl anion ring closure of 72, 73

Cyclooctatetraene dianions from 1,5-cyclooctadiene 60
from cyclooctatetraenes 25
oxidation to cyclooctatetraenes 34, 66
structures of 8
Cyclooctatetraenes from cyclooctatetraene dianions 34, 66
reactions with carbanions 34
reduction with alkali metals 25
Cyclopentadienes acidity of 17, 18, 20, 21
from ring openings 61, 62, 74
metallation of 23, 74
Cyclopentadienyl anions aromaticity of 18
from oxidative couplings 65
from ring openings 61, 62, 74
reaction with chlorocarbene 40
structures of 8
Cyclopentenolates structure of 10
2-Cyclopentenones from carbanions 44
Cyclopentenyl anions ring openings of 62, 72–74
Cyclopentyl anions ring openings of 61
Cyclophanes from dianion oxidations 66
Cyclopropanes acidity of 19
from carbanions 48, 57, 69, 73
metallation of 19
Cyclopropyl anions reactions without opening 72
ring opening of 72
structures of 7, 8

Darzens condensations 22, 46
DBU as a base for carbanion preparations 23
Dehydrogenation 65
Delocalization of charge extra importance in gas phase acidities 20
Desulfurizations 31, 54
Deuterations via carbanions 22, 29
Deuterium oxide reaction with carbanions 29
Dialkyl copper-lithium reagents see lithium dialkylcuprates
Dialkylchlorophosphines reactions with carbanions 30, 33
Dialkylmagnesiums 6, 14

N,N-Dialkyl-O-arenesulfonylhydroxyl-
amines reactions with carbanions
30, 33
Diammonium croconate 10
Diarylmagnesiums 14
1,8-Diazabicyclo[5.4.0]-7-undecene as a
base for carbanion preparations 23
1,2-Dibromoethane 14, 65, 66
Dicarbanions addition reactions of
50–52
eliminations from 57–61, 73
enediolate 50
insolubility of salts of 43
oxidations of 64–66
preparations of 12, 16, 18, 23–25,
27, 61, 65
reactions with dihalides 51, 65
rearrangements of 67–69, 73
sites of reactions of electrophiles with
42, 43
N-stabilized reactions of 55–57, 61
O-stabilized reactions of 50–52, 55,
56
S-stabilized reactions of 56
structures of 8–11
substitution reactions of 31, 49–52,
58
use of indicators 25
1,2-Dichloroethane 65
Dieckmann condensations 22, 23, 45
1,3-Dienes 1,4-addition of carbenes to
40
from protonation of pentadienyl anions 41, 42
isomerization of 42
metallation of 24, 25
nucleophilic additions to 34
reduction by Mg 25
1,4-Dienes from protonation of pentadienyl anions 41, 42
metallation of 24
1,5-Dienes from oxidative coupling of
allyl anions 64
Dienolate anions cross-conjugated reactions with electrophiles 48
from ring openings 72, 73
high rotation barriers in 72
linear favored α-attack in 40, 41,
47
linear favored γ-attack with CuX in
41, 48
ring closings of 73

Diethyl malonate acidity of 21
in malonic ester syntheses 48, 49,
51
Diethylmagnesium 6
Diglyme 8, 22
Dihalides reactions with dicarbanions
30, 33, 51, 65, 66
Diisobutyl ketone additions of carbanions to 42
Diisopropylamine conversion to LDA
23
1,3-Diketone enolates reactions with
electrophiles 48, 49
structures of 10, 11
1,2-Diketones from dianion oxidations
65
1,3-Diketones dimetallation of 51
metallation of 23, 48, 49
synthesis of 45, 49, 51
1,4-Diketones aldol condensations of
53
from carbanions 53
from oxidative additions 64
1,5-Diketones aldol condensations of
53
from carbanions 53
Dimers 5, 7, 9, 11, 43
oxidative 64–66
1,2-Dimethoxyethane 11, 14, 43
2,6-Dimethoxyphenyllithium 7
Dimethyl disulfide reactions with carbanions 61
Dimethyl malonate acidity of 17, 21
Dimethyl sulfate 14
as a hard electrophile in reactions
with enolates 43
Dimethyl sulfide 17
N,N-Dimethylacetamide acidity of 21
Dimethylberyllium 6
2,3-Dimethylenebutadiene dianion reaction with t-butyl bromide 31
N,N-Dimethylformamide 22, 25, 32,
43, 61
Dimethylmagnesium 3, 6
Dimethylsulfone acidity of 17, 21
Dimethylsulfoxide 9, 17, 21–23, 42,
43
acidity of 17, 21
Dimsyl anion as a base for carbanion
preparations 23
1,2-Dinitro compounds from oxidative
couplings 64

Dioxane 14
Dioxolane 11
Diphenylmethane acidity of 17, 21
Diphenyl(methyl)silyl chloride as a soft electrophile 43
Diradicals 66, 70
Disulfides reactions with carbanions 30, 33, 61
1,3-Dithiacyclohexane acidity of 17
1,3-Dithiacyclohexane-1,1,3,3-tetroxide acidity of 17
Dithioacetals acidity of 17
anions from 34, 61, 70
from carbanions and thioesters 30, 38
hydrolysis to aldehydes and ketones 31
DME 11, 14, 43
DMF 22, 43
formylations of anions with 25, 32, 61
DMSO 9, 17, 21–23, 42, 43
Donor atoms 5

Eglinton reaction 64
Electron affinities of radicals 20
Electron density in controlling attack of electrophiles on allyl anions 40
in organometallic bonds 3
Eliminations from carbanions 57–62
α- 57, 58
β- 58–60
cyclo- 37, 57, 61, 62
δ- 61
γ- 37, 46, 48, 60
of alkoxides 57, 58
of alkyl anions 34
of amide ions 57, 58
of carboxylates 45, 46, 49, 58, 59
of cyanide ion 58
of halide ions 57–60
of hydride ions 34, 59, 60
of hydroxide ion 45–47, 58
of mercaptide ions 61
of oxide ion 57, 58
of sulfinates 35, 60, 61
of trimethylsiloxide 45, 46, 59
Enaminates additions to aldehydes and ketones 45
alkylations of 44, 52–54
preparations of 26
structures of 10

Enamines alkylation of 44
from ring openings of nitrogen ylides 61
Enediolate anions oxidation to 1,2-diketones 65
reactions with CS_2 50
Enol phosphates preparation and reactions with lithium dialkylcuprates 45
Enolates acylations of 44, 45
additions of 22, 37, 45–47, 68
aggregation into dimers and tetramers 10–12, 43
aluminium 44
arylation of 44
as intermediates in aldol condensations 22, 68
boron 46
charge distribution in 10
eliminations from 58
from aldehydes 44
from alkylations of dianions 58
from esters 43–46
from isobutyrophenone structure of 11
from β-ketoesters reactions with electrophiles 49
from malonic esters reactions with electrophiles 49
kinetic vs. thermodynamic generation of 20
lithium 46
oxidations of 64, 66
O-phosphorylation of 45
polyalkylation of 44
preparations of 22, 23, 26, 32–34, 36, 43, 58, 61
protonations of 36
reactions at carbon vs. oxygen 43
regiospecific generation of 28
stabilization by alkyl groups of 18
stereochemistry of 46, 47
structures of 10–12
substitution reactions of 43–45, 60
tin 44
tris(dimethylamino)sulfonium 46
vinylation of 44
zinc 46
zirconium 46
Episulfides reactions with carbanions 30, 31
Epoxides conversions to oxetanes 37

from γ-eliminations 37, 46, 47
from oxidative couplings 66
reactions with carbanions 30, 31, 38, 51, 54, 69
reactions with lithium 58
silyl metallation of 37
Equilibrations *see* Thermodynamic control
Erythro configurations 46
Esters acidity of 18
acylation of 23, 45
alkylation of 23
α-epoxy- from Darzens condensations 46
from carbanions and carbonates 33
from malonic half ester dianions 52
β-keto- dimetallations of 51
from carbanions 45, 49, 51
hydrolyses and decarboxylations of 49
metallations of 18, 23, 49
reactions with carbanions 30, 32, 38, 39, 49, 55
succinic from carbanions 51
α,β-unsaturated from dianions 56
from dienolates and alkyl halides 48
from enolates and aldehydes or ketones 45, 46
from β-ketoesters 49
reductive acylations of 25
β,γ-unsaturated from dienolates and alkyl halides 44, 48
Ethanol acidity of 21
Ethers as solvents for carbanions 22, 29
cleavages by metals of 14, 15
cleavages by organometallics of 24
metallations of 18–20, 70
vinyl from alkylations of enolates 42–45, 52
from Wittig-type reactions 37
hydrolyses to ketones of 52, 53
Ethyl acetate acidity of 17
Ethyl chloroformate reactions with dianions 56
Ethyl hydrogen malonate in malonic ester syntheses 51
Ethylbenzene acidity of 21
from carbanion eliminations 60
N-Ethyl-4,6-dimethyl-2-oxidopyridinium as a leaving group 31

Ethylene, anionic polymerization of 34
from carbanion eliminations 60, 61
from cyclic sulfone eliminations 66
from hydride eliminations 59
Ethylene oxide preparations from carbanions 37
reactions with carbanions 51
reactions with lithium 58
Ethyllithium elimination of hydride from 59
structure of 3, 5, 6
Ethylmagnesium bromide 3, 5, 7
Ethylphenylsulfone acidity of 17
Ethylpotassium elmination of hydride from 59
from n-butylpotassium 60
Ethynylpotassium 7
Ethynylrubidium 7
Ethynylsodium 7
Exchange between aggregated carbanion forms 7

Ferric ion oxidations of carbanions by 64, 65
Ferrocenes open-chain 40
Ferrous chloride reactions with pentadienyl anions 40
Flowing afterglow 20
Fluorene acidity of 17, 20, 21
Fluorenyl anions 9
Fluorenyllithium 8
Fluorenylpotassium 8
Fluoroform acidity of 21
Formaldehyde reactions with dianions 50–52, 56
Formates reactions with carbanions 33
Fulvene dianion oxidation of 65
Furans metallations of 19
ring cleavage of metallated 70

Gas phase carbanions 3, 5, 6, 8, 9, 45
Glycidic esters preparations of 46, 66
Glyme 11, 14, 43
Grignard reagents mechanism of formation of 14
preparations of 14, 36
reactions with: acetylenic sulfones 35
aldehydes and ketones 36, 37
alkyl halides 31, 38
amides 32

anhydrides 32
N,N-dialkyl-O-arenesulfonyl-
hydroxylamines 30, 33
D_2O 29
epoxides 31
imines 40
iron pentacarbonyl 33
propargyl acetals 31
thioesters 32
α,β-unsaturated carbonyls 34
vinyl halides 32
reductions of ketones by 59
structures of 3–5, 7, 14

α-Haloesters reactions with zinc 15
Halogen exchange of aryl and vinyl
halides 32
Halogens reactions with carbanions of
30, 33, 64, 66
Hard vs. soft electrophiles and nucleophiles 43, 44
Heptafulvene dianion from 1-methyl-
cycloheptene 60
Heptatrienyl anions electrocyclizations
of 72, 74
failure to undergo [1,8]sigmatropic
proton shifts of 71
from sigmatropic rearrangements
74
from tetraenes 34
metallations of 34
rotation barriers in 72
Heteroatoms in carbanions 19, 20, 31,
43–58, 60–62, 64–66, 68–73
removal of 31
Hexachloroethane reactions with carbanions 30, 33
Hexamers 6
Hexamethylphosphoramide 22, 24, 32
Hexatriene dianions electrocyclizations
of 73
hetero analogs of 55, 56
structure of 8
HMPA 22, 24, 32
Homoaromaticity in carbanion stabilization 18
Homocyclooctatetraene dianions 18
Homocyclopentadienyl anions 18
Homolytic dissociation of carbon acids
20
Horner-Emmons reactions 37

Hybridization sp 7
sp^2 7, 8, 12
sp^3 3–7
Hydrazones, acylation and alkylation
of dimetallated 55
conversion to pyrazoles 55
dimetallation of 55
from alkylation of metallated hydrazones 53
metallation of 53
Hydride eliminations from carbanions
57, 59, 60, 73
Hydride transfers from alkoxides 22
from carbanions 36, 59, 60
Hydrochloric acid acidity of 21
Hydrofluoric acid acidity of 21
Hydrogen acidity of 17, 21
Hydrogen cyanide from cyanohydrins
64
Hydrogen peroxide in oxidizing alkylboranes to alcohols 41
Hydrogen sulfide acidity of 21
Hydroperoxides from carbanions 63,
64
reductions to alcohols of 63, 64
Hydrostannation of alkynes 28
Hydroxide ion as a base for carbanion
preparation 21, 22
as a leaving group in eliminations
45–47, 58
Hyperconjugation 9, 18

Imidazoles dihydro- from carbanions
73
Imines additions of carbanions to 30,
35, 36, 40, 47
as intermediates in transaminations
42
from oxidative coupling of enaminates 64
from reactions of enaminates with
electrophiles 52, 53
hydrolyses 36, 52, 53
instability of 35
α,β-unsaturated additions of carbanions to 35
Iminium salts additions of carbanions
to 30, 36, 47
from aldehydes and secondary amines
47
Indene acidity of 17
Indenyllithium 8

bis-Indenylmagnesium 8
Indoles from enolate ions and *o*-halo-anilines 44
Inductive effects in charge stabilization 18, 19, 34
Infrared studies 7, 9
Iodine oxidations of carbanions by 64, 66
Iodobenzene reaction with methyllithium 32
Ion cyclotron resonance 20
Ion pairs 8, 9, 11, 67
Ionization potential of a hydrogen atom 20
Iron pentacarbonyl, reaction with Grignard reagents 33
Isobutylene mono- and dimetallation of 24
Isocyanides from ring openings of metallated isoxazoles 69
  α-metallated additions to aldehydes and ketones of 37
  metallations of 18, 19
Isomerizations *see* also Rearrangements
  geometric 7, 8, 67
  optical in carbanions 7, 8, 67
  positional in alkenes 22, 41, 42, 67
    in alkynes and allenes 68
    in allyl anions 67
    in dianions 67, 68
Isonitriles *see* Isocyanides
Isopropylbenzene acidity of 21
Isoxazoles from carbanions 55
  metallated ring opening of 61, 69
Isoxazolidones from carbanions 55

Journal of Organometallic Chemistry 1

Ketones additions of carbanions to 28, 30, 36, 37, 40, 42, 45–47, 49, 56, 68
  additions of enolates to 45–47
  α-amino- from carbanions 56 47
  aryl from carbanions 71
  aryl, from carbanions 71
  by acylation of carbanions 32, 33
  by hydrolysis of imines 36, 53
  cyclopropyl from enolates 60
  diaryl electron transfer to 36

enolization of 36
  from carbanions and carboxylates 38
  from carbanions *via* dithioacetals 30, 31, 38
  from carboxylic acid dianions 63, 64
  from chiral alkylations 53, 54
  from β-ketoesters 49
  from α-metallated nitriles 63, 64, 71
  from nitrate ions 63, 64
  from α-silylepoxides 37
  α-halo- reactions with carbanions 58
  β-hydroxy- elimination of water from 45–47
    from aldol condensations 45–47
    from dienolates 48
  metallations of 23
  reductions to alcohols of 36, 59
  α,β-unsaturated from aldol condensations 45–47, 53
    from α,γ-diketoesters 49
    from α-metallated-α,β-unsaturated ketone equivalents 48
    nucleophilic additions to 48
  β,γ-unsaturated from carbanions 51, 71
Kinetics 16, *see* also Rate control

Lactones β- from carbanions 50
  α-keto-γ- pyrolysis of 49
  preparations of 40, 48, 52
  α,β-unsaturated-γ- preparations of 52
LAH in ketone reductions 50
LDA as a base for carbanion preparations 23, 28, 42
Leaving groups in eliminations from carbanions 57
  in nucleophilic substitutions 29, 31
Lithiocarbons 10
α-Lithioethers preparations of 15
2-Lithio-2-methyl-1,3-dithiane 5
α-Lithiosulfoxides 12
Lithium reactions with alkyl halides 13, 15, 16, 36, 58
  reactions with epoxides 58
  reactions with terminal alkenes 60
  transmetallations with 27

Lithium amide as a base for carbanion preparations 23

Lithium amidines as bases for carbanion preparations 23

Lithium *t*-butoxide as a base for alkene isomerizations 22
insolubility in hexane 24

Lithium carbonate 31

Lithium carboxylates reactions with alkyllithiums 32

Lithium dialkylcuprates 1,4-additions to α,β-unsaturated carbonyl compounds of 34
from vinyl anions 38
mixed to save alkyl groups 29, 34
preparations of 29, 38
reactions with: alkyl halides 29
enol phosphates 45
epoxides 31
vinyl halides 32

Lithium dicyclohexylamide as a base for carbanion preparations 23

Lithium diisopropylamide as a base for. carbanion preparations 23

Lithium fluoride 31

Lithium hexamethyldisilamide as a base for carbanion preparations 23, 50

Lithium hydride, eliminations of 59, 60

Lithium naphthalenide 15

Lithium oxide eliminations of 58
structure of 7

Lithium 2,2,6,6-tetramethylpiperidide as a base for carbanion preparations 23

LiTMP as a base for carbanion preparations 23

Macrolide antibiotics, synthesis of 46

Magnesium reactions with alkyl halides 13, 15, 36
reductions of conjugated dienes with 25

Magnesium ion as a counterion for dianions 51, 56
in controlling epoxide ring openings 69

Malonic ester syntheses 22, 48, 49, 51

Malononitrile, acidity of 17, 21

Mannich reactions 47

Mass spectrometry high pressure 20

Mechanisms addition-elimination 32, 35, 59, 70, 73
benzyne 32
ElcB 58
elimination-addition 32, 74
of alkyl halide reactions with organo-metallics 29, 31
oxidation 63
SET 31, 33, 36, 63, 65, 66
S_N1 32, 43
S_N2 29, 31, 32, 43, 63
S_N2 31, 53
$S_{RN}1$ 31, 44

Meisenheimer reactions 26

Menthyllithium 5, 20

Mercuric acetate in mercuriation of alkenes 27

Metal-halogen interchanges 15

Metallations 17–21
kinetic *vs.* thermodynamic 20, 24, 67, 68
of alkenes 19
of alkyl halides 18, 19
of alkynes 19
of amides 18, 20
of amines 18–20
of aromatics 19, 20
of carboxylic acids 20
of cyclopropanes 19
of esters 18
of ethers 18–20
of furans 19
of isonitriles 18, 19
of ketones 20
of N-methylpyrrole 19
of phosphine oxides 18
of phosphines 18
of phosphonates 19
of phosphonium salts 18
of pyridines 19
of pyrroles 19
of selenides 18
of silanes 18, 19
of sulfides 18, 19
of suflonamides 20
of sulfones 19
of sulfonium salts 18
of sulfoxides 19
of thiophenes 19
ortho- 19–21

Methane acidity of 17, 21

Methanol acidity of 17, 20, 21

Methyl acetate acidity of 21
Methyl carbanion from eliminations 60
  structure of 3, 6
Methyl iodide as a soft electrophile in reactions with enolates 43
  oxidations of dianions by 66
  reactions with: dithiocarbamates 31
    dithiocarbonates 50
    dithiocarboxylates 51
    enaminates 53
    thioenolates 56
N-Methyl phenyl methyl sulfoximine acidity of 17
Methyl phenyl sulfone acidity of 17
Methylacetylene 10
2-Methyl-2-butene dimetallation of 68
Methylcesium 6
1-Methylcyclopropene isomerization of 42
Methylene halides reactions with carbanions 30, 33
2-Methyleneallyl dianions oxidations of 66
  preparations of 68
Methylenecyclopropane preparations of 42, 66
Methyllithium cleavage of silyl ethers with 28
  reaction with D₂O 29
  reactions with iodobenzene 32
  structure of 3, 5, 6
Methylmagnesium iodide reaction with D₂O 29
Methylmalonitrile, acidity of 17
Methylpotassium 6
N-Methyl-2-pyridyl formates reactions with Grignard reagents 32
N-Methylpyrrole metallations of 19
N-Methylpyrrolidone 22
Methylrubidium 6
Michael additions 26
Molecular orbital calculations on carbanions 3, 9, 10, 18, 21
Molybdenum complexes, oxidations of carbanions with 63
Monomers 3, 7
MSAD acidity scale 17
Myrcene reduction with Mg 25

Naphthalene reduction to anion radical 25

Naphthalene dianion preparation of 25
  structure of 8
Natural product syntheses involving carbanions 26, 46
NBS reactions with carbanions 30, 33, 38
NCS reactions with carbanions 30, 33, 39
Neutron diffraction 7
Nickel halides as catalysts for reactions of carbanions with aryl and vinyl halides 32, 44
Nitriles see also Acetonitriles
  additions of carbanions to 25, 30, 36
  α-amino- conversions to aldehydes 54
    preparations from carbanions 54
  by nucleophilic additions to α,β-unsaturated nitriles 35
  eliminations of cyanide ion from 58
  from carbanions and tosyl cyanide 33
  from ring openings of metallated isoxazoles 61, 69
  from Wittig-type reactions 37
Nitroarenes from carbanions and alkyl nitrates 30, 33
  nucleophilic additions to 26
  oxidations of carbanions with 66
Nitroalkanes acidity of 18, 20
  from carbanions and nitroalkenes 35
  metallation to nitroates 18, 20, 32, 49, 64
  reduction by copper salts
1-Nitroalkenes nucleophilic additions to 35
Nitroates from dianions 52
  from nitro compounds 18, 20, 32, 49, 64
  reaction with singlet oxygen 64
  stabilization by α-alkyl groups 18
Nitrobenzene see Nitroarenes
Nitroethane acidity of 17, 21
Nitromethane acidity of 17, 21
2-Nitropropane acidity of 21
N-Nitrosoamines alkylations of 54
  denitrosations of 54
  metallations of 54
p-Nitrotoluene acidity of 17, 21

NMP   22
Nuclear magnetic resonance   3, 9–12, 18
^{13}C   9, 11
^{7}Li   9

Organoaluminiums reactions with allenyl halides   31
structures of   3–5, 8
Organoberylliums   4, 6
Organocadmiums reactions with acid chlorides   33
Organocesiums   4, 6
Organocuprates   26, 33, 40, 41 see also Lithium dialkylcuprates
Organolead compounds transmetallation of   27
Organolithiums preparations of   15, 27
reactions of see Carbanions, reactions structures of   3–13
Organomagnesiums see Grignard reagents and Dialkylmagnesiums
Organomanganese reagents, reactions with acid chlorides   33
Organomercury compounds transmetallation of   27
Organometallics see Carbanions
Organopotassiums low solubility of   24
metallations with   24
structures of   3, 4, 6
use in preserving stereochemistry   24
Organorhodiums reactions with acid chlorides   33
Organorubidiums   3, 4, 6
Organosilicon compounds see also Trimethylsilyl chloride
transmetallation of   27
Organosodiums metallations with   24
structures of   4
Organosulfur compounds desulfurization of   31
Organotin compounds in alkene-forming eliminations   37
transmetallation of   27, 28
Organozincs, acetylenic   38
as Reformatsky intermediates   15
from Rieke zinc   15
Ortho acylation of arenes   71
Oxazines, from metallated oxazines   54
hydrolysis to acids   54

metallation of   54
Oxazoles, dihydro- from isocyanides   37
metallated ring openings of   69
Oxazolines from metallated oxazolines   54
hydrolysis to acids   54
metallation of   54
Oxetanes from epoxides   37
α-metallated   37
Oxidations of carbanions   34, 63–66
Oxidative couplings of carbanions   33, 64–66
mixed   64
Oxide ion eliminations from carbanions of   57, 58
2-Oxidopyridine as a leaving group   31
Oxygen reactions with carbanions   30, 33, 63, 64, 66
singlet, reactions with carbanions   64
Palladium catalysts   32, 38
1,3-Pentadiene reduction to anion radical   25
1,4-Pentadiene metallation of   24
Pentadienyl anions as intermediates in nucleophilic aromatic substitutions   32, 59
boronation of   41
control of shape of   41, 72
2,4-diaza-   73
eliminations from   32, 59, 60
equilibrations with cyclopentenyl anions   72–74
metallations of   34
oxidative couplings of   64
planar shapes of   9, 41, 67, 72
position of attack by electrophiles in   40–42, 65
preparations of   24–26, 34, 41, 59, 74
reactions with FeCl$_2$ 40
[1,6]sigmatropic proton shifts in   70, 74
structures of   9

Peterson reactions   37
Phase transfer systems for Darzens and Stobbe condensations   22
Phenol, acidity of   17, 21
Phenolates   11

Phenyl arenesulfonates reactions with carbanions 30, 33
Phenyl methyl sulfide acidity of 17
Phenylacetonitrile, acidity of 17, 21
Phenylacetylene acidity of 17
Phenyllithium 7
Phenylmagnesium bromide 7
Phenylmalononitrile acidity of 17
Phenylpotassium 15
Phenylsodium 15
Phosphates as promoters of Grignard reactions with alkyl halides 31
Phosphine oxides, as Wittig products 37
  metallations of 18
Phosphines metallations of 18
  preparations from carbanions and chlorophosphines 30, 33
  use to save alkyl groups in cuprates 34
Phosphites, trialkyl desulfurizations with 50
  reductions of hydroperoxides by 63
Phosphonates from diethyl chlorophosphate 54
  in Horner-Emmons reactions 37
  metallations of 19
Phosphonium salts in Wittig reactions 37
  metallations of 18
Phosphoryl chlorides reactions with enolates 45
Photochemical carbanion reactions 44, 67, 70
Phthalocyanine dianion dipotassium 8
Polarization in charge stabilization 18
Polycarbanions 19, 23
Polymeric organometallics 6, 7, 10, 11, 53
Polymerization anionic 26, 34
  of p-xylylene 66
Polymetallation 19, 23
Potassium metallations with 23
  reduction of 1,1,2,2,-tetraphenylethane with 26
  transmetallations with 27
Potassium amide as a base for carbanion preparations 23
Potassium t-butoxide as a base for carbanion preparations 22, 24, 42

positional isomerization of alkenes with 41, 42
Potassium hydride as a base for carbanion preparations 23
  elimination from carbanions of 59
Precipitation as a driving force for carbanion formation 24, 27
Prenyl anions additions to ketones of 42
  regioselective alkylations of 41
Preparations of carbanions 13–28, 67–74
Propargyl acetals conversions to allenyl ethers 31
Propargyl halides reactions with carbanions 30, 31
Propene acidity of 17, 21
  metallation of 24
Propiophenone acidity of 17
Propylene oxide reaction with carbanions 54
1-Propynylpotassium 7
1-Propynylsodium 7
Protecting groups carbanions as 50
Proton affinities of carbanions in the gas phase 20, 21
Pseudoequatorial substituents 46, 47
Pyrazines from carbanions 55
Pyridines additions of nucleophiles to 26, 36
  from nucleophilic additions to pyridium salts 47
  metallation of 19, 35
Pyridinium salts additions of carbanions to 47
α-Pyridones thio-, from carbanions 50
S-2-Pyridyl thioates reactions with Grignard reagents 32
Pyrroles, metallation of 19
Pyrrolidine acidity of 17

Quaternary ammonium hydroxides as bases for carbanion preparations 22
Quinones from oxidation of hydroquinone dianions 65
Quinuclidine 8

Racemization of optically active carbanions 67
Radical intermediates, combination with anions to give anion radicals 44

coupling of 16, 31, 63, 66
disproportionation of 31
from carbanions by electron transfer
   36, 63
from cleavages of anion radicals 26,
   31, 44
in Grignard reagent formation 14
in reactions of alkylithiums with alkyl
   and aryl halides 16
reduction to carbanions 26
Raman studies 9
Ramberg-Bäcklund reactions 60
Raney nickel desulfurizations 31, 54
Rate control in additions 34, 42, 46,
   51
   in aldol condensations 53
   in forming 5-rather than 6-membered
      rings 68
   in metallations 24, 67, 68
   in protonations of allyl anions 40,
      41
   in protonations of pentadienyl anions
      41, 42
Rates of ionization 20
Rearrangements carbanion 13, 16,
   67–74
   complex 73, 74
   electrocyclic 72–74
   intermolecular 67, 68, 74
   intramolecular 68–74
   of alkyl groups from boron to carbon
      in alkylboranes 29
   sigmatropic 70, 71, 74
Recycling in isomerizations 42
Reductions electrochemical 25
   of σ bonds 25
   one-electron 25
   to form dicarbanions 25
   two-electron 25
   with active metals 25
Reductive acylations of alkene derivatives 25
Reformatsky reactions 15
Reimer-Tiemann reactions 57
Resonance in stabilizing carbanions 8,
   18, 34, 35
Rieke magnesium 14
Ring currents in cyclooctatetraene dianions 18
Ring strain in 3-and 4-membered rings
   69, 71–74

Rotation barriers in allyl anions 9, 10,
   63, 67
   in enolate ions 9, 10
   in heptatrienyl anions 72
   in pentadienyl anions 9, 72
   in vinyl anions 7, 8
Rotation mechanisms in π carbanions
   9, 10, 67

Salt effects 16
Sandwich structures 9, 40
Schlenk equilibrium 14
Selenides metallations of 18
   oxidation to selenoxides 31
Selenoxides preparations from selenides
   31
   pyrolysis to alkenes 31
Sesquiacetylenic dianions 10
SET see Single electron transfer
Shapiro reaction 57, 61
Silanes from carbanions and silyl hali-
   des 33, 43
   metallation of 18, 19
   vinyl- additions to
      from Wittig-type reactions 37
Silicon-oxygen bond cleavage by orga-
   nometallics 16
Single electron transfer, from carb-
   anions to:
      aldehydes and ketones 36
      alkyl halides 31
      oxidizing agents 63
      oxygen 63
      silyl halides 33
   from dianions 65, 66
   to radicals from superoxide 63
Sodium, metallations with 23
   reactions with alkyl halides 13, 15
Sodium amide as a base for carbanion
   preparations 23
Sodium hydride as a base for carbanion
   preparations 23, 44
Sodium sulfite reduction of hydroperoxides with 63
Solubility of carbanions 9
Solvation energies in stabilizing carbanions in solution 20, 21
Solvents cation-solvating 9, 22, 43
   coordinating 3, 24
   ether 3, 5, 7, 14–16, 24, 43
   for carbanion preparations 22–25
   hydrocarbon 5, 6, 8, 10, 15, 16, 24

polarity of in C *vs.* O alkylation of enolates, 43
tertiary amine 3, 23, 24
Spectroscopy IR and Raman 7, 9
  NMR 3, 9–12, 18
  photoelectron 3
  UV-visible 9
Stabilizing α-groups in alkenes 40, 41
  in allyl anions 40, 41
  in carbanions 18, 19
Stannous ion reduction of hydroperoxides with 64
Steric effects in additions 35, 40
  in sigmatropic rearrangements 70
  on carbanion stability 18
Stevens rearrangements 70
Stobbe condensations 22
Strontium reactions with alkyl halides 15
Structures of carbanions 3–12
Styrenes, additions of carbanions to 26, 51
  reductive acylation of 25
Substitution reactions of acetylenic, aryl, and vinyl anions 38, 39
  of alkyl anions 29–33
  of aryl halides 59
Succinimide acidity of 21
Sulfates reactions with carbanions 29
Sulfides from carbanions 30, 33
  from sigmatropic rearrangements 70
  metallation of 18, 19
  oxidation to sulfoxides 31
  vinyl, additions to 34
    from dianions 56, 61
  hydrolysis to aldehydes 61
Sulfinates as leaving groups 35, 60, 61
Sulfonates reactions with carbanions 29
Sulfones acidity of 18
  cyclic from Ramberg-Bäcklund reactions 60
  dimetallations of 66
  eliminations from 35, 60, 66
  from carbanions and arenesulfonates 30, 33
  from oxidations of dicarbanions 66
  metallations of 19, 60, 69
  vinyl from enolates 45, 46
    from Wittig-type reactions 37

Sulfonium salts eliminations from 37
  metallations of 18
  vinyl additions of carbanions to 47
Sulfonyl chlorides reactions with carbanions 30, 33
Sulfoxides acidity of 18
  from sulfides 31
  α-metallated additions to aldehydes and ketones 37
  metallations of 19
  pyrolyses to alkenes 31, 37
Sulfur reactions with carbanions 30, 33
Sulfur dioxide eliminations of 66
Syn additions of carbanions to alkynes 35

Tautomerizations *see* Isomerizations
Telomerization of butadiene 34
Temperature effects in rate *vs.* thermodynamic control 46, 51, 53
  on carbanion aggregation 43
  of epoxide ring openings 69
Terpene synthesis, butadiene dianions in 25
Tetraalkylammonium fluoride 28
Tetraalkylammonium salts of carbanions 28, 49
Tetrahydrofuran 7, 9–11, 14, 16, 43
  α-metallated cleavage of 61
Tetramers 5, 6, 11, 43
Tetramethylethylenediamine as an activating agent for organometallics 24
  chiral analogs of in asymmetric syntheses 36
  structure in organometallic complexes 3, 5–8
1,1,2,2,-Tetraphenylethane reduction with K 26
Thermal interconversions of organometallics 10
Thermodynamic control, in additions 34, 42, 46, 48, 51
  in aldol condensations 53, 68
  in alkene isomerizations 41, 67
  in alkyne and allene isomerizations 68
  in allyl anions 67
  in dianions 67, 68
  in metallations 24
  variation with temperature 42

THF 7, 9–11, 14, 16, 43
Thiazoles desulfurization of 54
  from metallated thiazoles 54
  hydrolysis to acids 54
  metallation of 54
Thioacids from carbanions and CS_2 30, 38
  from ring openings of dithioacetal anions 61
  from sigmatropic rearrangements 71
Thioamides from metallated thioamides 56
  metallated reactions with electrophiles 56
Thiocarbamates conversions to alkenes 31
Thiocarbonates from enediolate dianions 50
  pyrolyses of 50
Thiocyanates reactions with carbanions 30, 33
Thioesters additions of carbanions to 38
  from carbanions 51, 56
  in alkene-forming eliminations 37
  metallated 56, 71
Thiophenes from dianion oxidations 66
  metallations of 19, 66
  ring openings of side-chain metallated 70
Thiourea acidity of 17
Threo configurations 46
TMEDA see Tetramethylethylenediamine
TMSCl see Trimethylsilyl chloride
Toluene acidity of 17, 20, 21
  from methyllithium and iodobenzene 32
Tosyl chloride reactions with carbanions 30, 33
Tosyl cyanide reactions with carbanions 33
Transaminations 41
Transition metals complexes of as catalysts for carbanion reactions 44
Transmetallations 27, 30, 33
Trapping experiments 14
Tributyltin chloride reactions with enolates 44

Tributyltin hydride additions to alkenes of 28
Tricarbanions, eliminations from 59, 60
  preparations of 34
  reactions of electrophiles with 42, 43
  use in synthesis 49
Trichloroethene reactions with dienolate ions 48
Tricyclohexene preparation of 40
Trienolate ions from ring openings 72, 73
  high rotation barriers in 72
  reactions with electrophiles 48
  use as protecting groups for dienones 50
Triethanolamine borate in alkylation of enolates 44
Triethylaluminium, reactions with enolates 44
Triethylamine 5
Triethylenediamine 8
Trifluoroacetic acid acidity of 21
1,1,1,-Trihalides hydrolyses to carboxylic acids 31
Trimethylaluminium 5
Trimethylsilanol elimination of 45, 46
Trimethylsilyl chloride as a hard electrophile 43
  in trimethylsilylations of carbanions 28, 33, 43, 48, 54
  in trimethylsilylations of other anions 52, 54

Trimethylsilyl vinyl ethers from enolates 48
  in regiospecific generation of enolates 28
Trimethylsilylmethylpotassium as a superior metallating agent 24
  preparation by transmetallation 27
  reactions with esters 33
Trinitromethane acidity of 21
Triphenylmethane acidity of 17
Triphenylmethyllithium 8
Triphenylmethylsodium 8
Triple ion structure 9
Tris(diethylamino)sulfonium cations in enolate reactions 46
Tris(phenylthio)methane acidity of 17

Ultrasounds 14
Ultraviolet-visible studies 9
Unshared pairs coordination of metals
  with 11
Urea acidity of 17

Vinyl anions configurations of 35, 38
  eliminations from 60, 61, 69, 73
  isomerizations of 35
  preparations of 13, 16, 24, 32, 35,
    55, 57, 61, 73
  reactions with electrophiles 32, 35,
    38, 52, 55, 61
  structures of 7
Vinyl halides from vinyl anions and
  NBS 38
  reactions with carbanions 30, 32, 44
  reactions with strong bases 60
  reactions with transition metal com-
    pounds 32, 44
Vinyl selenides additions to 34
Vinyl arsines additions to 34
Vinylcopper reagents preparations of
  35
  reactions with electrophiles 35, 38
Vinyllithiums preparations of 16, 27,
  28

Vinyltin compounds as carbanion equivalents 28

Water acidity of 17, 20, 21
  as a solvent for carbanions 22
Wittig reactions 37
Wittig rearrangements 70
Woodward-Hoffmann rules 70
Wurtz reactions 13–15, 29

X-ray studies 3–8, 10, 11
o-Xylylene preparation and polymerization of 66
p-Xylylene preparation and polymerization of 66

Ylides nitrogen 61, 70, 71
  sulfonium 47, 70, 71

Zinc, reactions with: alkyl halides 13
  α-halocarboxylates 15
  α-haloesters 15
Zinc ions in promoting 1,4-additions
  to α,β-unsaturated carbonyl compounds 35
  in making zinc enolates 46
  oxidative couplings of carbanions
    with 64

Reactivity and Structure

Concepts in Organic Chemistry

Editors: K. Hafner, J.-M. Lehn, C. W. Rees,
P. v. R. Schleyer, B. M. Trost, R. Zahradník

Volume 1: J. Tsuji
Organic Synthesis
by Means of Transition Metal Complexes
A Systematic Approach
1975. 4 tables. IX, 199 pages. ISBN 3-540-07227-6

Volume 2: K. Fukui
**Theory of Orientation
and Stereoselection**
1975. 72 figures, 2 tables. VII, 134 pages
ISBN 3-540-07426-0

Volume 3: H. Kwart, K. King
**d-Orbitals in the Chemistry of Silicon,
Phosphorus and Sulfur**
1977. 4 figures, 10 tables. VIII, 220 pages
ISBN 3-540-07953-X

Volume 4: W. P. Weber, G. W. Gokel
**Phase Transfer Catalysis
in Organic Synthesis**
1977. Out of print. New edition in preparation

Volume 5: N. D. Epiotis
Theory of Organic Reactions
1978. 69 figures, 47 tables. XIV, 290 pages
ISBN 3-540-08551-3

Volume 6: M. L. Bender, M. Komiyama
Cyclodextrin Chemistry
1978. 14 figures, 37 tables. X, 96 pages
ISBN 3-540-08577-7

Volume 7: D. I. Davies, M. J. Parrott
Free Radicals in Organic Synthesis
1978. 1 figures. XII, 169 pages
ISBN 3-540-08723-0

Volume 8: C. Birr
**Aspects of the
Merrifield Peptide Synthesis**
1978. 62 figures, 6 tables. VIII, 102 pages
ISBN 3-540-08872-5

Volume 9: J. R. Blackborow, D. Young
**Metal Vapour Synthesis in
Oranometallic Chemistry**
1979. 36 figures, 32 tables. XIII, 202 pages
ISBN 3-540-09330-3

Volume 10: J. Tsuji
**Organic Synthesis with
Palladium Compounds**
1980. 9 tables. XII, 207 pages. ISBN 3-540-09767-8

Volume 11
New Syntheses with Carbon Monoxide
Editor: J. Falbe
With contributions by H. Bahrmann, B. Cornils,
C. D. Frohning, A. Mullen
1980. 118 figures, 127 tables. XIV, 465 pages
ISBN 3-540-09674-4

Volume 12: J. Fabian, H. Hartmann
Light Absorption of Organic Colorants
Theoretical Treatment and Empirical Rules
1980. 76 figures, 48 tables. VIII, 245 pages
ISBN 3-540-09914-X

Volume 13: G. W. Gokel, S. H. Korzeniowski
Macrocyclic Polyether Syntheses
1982. 89 tables. XVIII, 410 pages
ISBN 3-540-11317-7

Volume 14: W. P. Weber
Silicon Reagents for Organic Synthesis
1983. XVIII, 430 pages. ISBN 3-540-11675-3

Volume 15: A. J. Kirby
**The Anomeric Effect and Related Stereo-
electronic Effects at Oxygen**
1983. 20 figures, 24 tables. VIII, 149 pages
ISBN 3-540-11684-2

Springer-Verlag
Berlin
Heidelberg
New York
Tokyo

H. J. Fischbeck, K. H. Fischbeck

Formulas, Facts and Constants

for Students and Professionals in Engineering, Chemistry and Physics

1982. XII, 251 pages. ISBN 3-540-11315-0

Contents: Basic mathematical facts and figures. – Units, conversion factors and constants. – Spectroscopy and atomic structure. – Basic wave mechanics. – Facts, figures and data useful in the laboratory.

This book provides a handy and convenient source of formulas, conversion factors and constants for students and professionals in engineering, chemistry, mathematics and physics. Section 1 covers the fundamental tools of mathematics needed in all areas of the physical sciences. Section 2 summarizes the SI system (International System of Units of measurement), lists conversion factors and gives precise values of fundamental constants. Sections 3 and 4 review the basic terms of spectroscopy, atomic structure and wave mechanics. These sections serve as a guide to the interpretation of modern literature. Section 5 is a resource for work in the laboratory, listing data and formulas needed in connection with frequently used equipment such as vacuum systems and electronic devices. Material constants and other data are listed for information and as an aid for estimates or problem solving.

Formulas and tables are accompanied by examples in all those cases where their use might not be self-explanatory.

Springer-Verlag
Berlin
Heidelberg
New York
Tokyo